Encyclopedia of Alternative and Renewable Energy: Biomass Processing and Production

Volume 11

Encyclopedia of Alternative and Renewable Energy: Biomass Processing and Production

Volume 11

Edited by **Brad Hill and**
David McCartney

New York

Published by Callisto Reference,
106 Park Avenue, Suite 200,
New York, NY 10016, USA
www.callistoreference.com

Encyclopedia of Alternative and Renewable Energy:
Biomass Processing and Production
Volume 11
Edited by Brad Hill and David McCartney

International Standard Book Number: 978-1-63239-185-8 (Hardback)

Printed in the United States of America

Contents

Preface

Over the recent decade, advancements and applications have progressed exponentially. This has led to the increased interest in this field and projects are being conducted to enhance knowledge. The main objective of this book is to present some of the critical challenges and provide insights into possible solutions. This book will answer the varied questions that arise in the field and also provide an increased scope for furthering studies.

Research in the field of biomass and related disciplines has increased with time. Awareness regarding pros and cons of biomass, sustainable use of resources and biomass sources has brought in the concept of biorefineries. This is evident from growth in biomass research and increasing attention towards biofuels. This book covers various disciplines of biomass under separate themes namely, biomass processing and biomass production.

I hope that this book, with its visionary approach, will be a valuable addition and will promote interest among readers. Each of the authors has provided their extraordinary competence in their specific fields by providing different perspectives as they come from diverse nations and regions. I thank them for their contributions.

Editor

Biomass Processing

Biomass Extraction Methods

Adina-Elena Segneanu, Florentina Cziple, Paulina Vlazan,
Paula Sfirloaga, Ioan Grozescu and Vasile Daniel Gherman

Additional information is available at the end of the chapter

1. Introduction

Biomass represents an extremely valuable potential to obtain new clean energy sources and natural structurally complex bioactive compounds. Renewable energy can be produced from any biological feedstock, that contains appreciable amounts of sugar or materials that can be converted into sugar (e.g. starch or cellulose). Lignocellulose's biomass–dendromass and phytomass is natural based material consisting of complex of heterogenic macromolecules with cell structure (celluloses, hemicelluloses and lignin) as well as numerous organic and inorganic structures with low molecule weight (Sun, 2002).

Long-term economic and environmental concerns have resulted in a great amount of research in the past couple of decades on renewable sources of liquid fuels to replace fossil fuels. Producing of cellulose and alcohol from biomass is important technological process. Conversion of abundant lignocellulosic biomass to biofuels as transportation fuels presents a viable option for improving energy security and reducing greenhouse emissions. Lignocellulosic materials such as agricultural residues (e.g., wheat straw, sugarcane bagasse, corn stover), forest products (hardwood and softwood), and dedicated crops (switchgrass, salix) are renewable sources of energy. These raw materials are sufficiently abundant and generate very low net greenhouse emissions. The use of biomass with low economic value, the waste from agriculture, forestry and wild flora as sources of clean energy, is a viable way to avoid potential conflicts with the biomass production for food, which represent the main concern of UE regarding the biofuels production from biomass.

The presence of lignin in lignocelluloses leads to a protective barrier that prevents plant cell destruction by fungi and bacteria for conversion to fuel. For the conversion of biomass to fuel, the cellulose and hemicellulose must be broken. The digestibility of cellulose present in lignocellulosic biomass is hindered by many physicochemical, structural, and compositional factors. The lignocellulosic biomasses need to be treated prior to fuel production to expose cellulose. In present, there is many different type of pretreatment of lignocelluloses materials. Pretreatment uses various techniques, including ammonia fiber explosion, chemical treatment, biological treatment, and steam explosion, to alter the structure of cellulosic biomass to make cellulose more accessible. The purpose of the pretreatment is to remove lignin and hemicellulose, reduce cellulose crystallinity, and increase the porosity of the materials. Then, acids or enzymes can be used to break down the cellulose into its constituent sugars. Enzyme hydrolysis is widely used to break down cellulose into its constituent sugars. Pretreatment can be the most expensive process in biomass-to-fuels conversion but it has great potential for improvements in efficiency and lowering of costs through further research and development. Cellulose chains can also be broken down into individual glucose sugar molecules by enzymes known as cellulose. Cellulose refers to a class of enzymes produced by fungi, bacteria, and protozoans that catalyze the hydrolysis of cellulose. But, one of the main drawn back of convention chemical methods used in ethanol formation process is degradation of carbohydrates and formation of undesirable by-products, which severely inhibition of ethanol during the fermentation process: furfural, 5-hydroxymethylfurfural, uronic acid, levulinic acid, acetic acid, formic acid, hydroxybenzoic acid, vanillin, phenol, cinnamaldehyde, formaldehyde, and so (Nenkova et.al., 2011). Some inhibitors such as terpene compounds are present in the biomass–dendromass.

Lignin is a complex reticulated phenolic polymer that occurs in xylem of most terrestrial plants and is the second most abundant biopolymer in nature, corresponding to around 30% of the biosphere organic carbon. This macromolecule is one of the biggest wood components and also one of the most important. Even the lignin has a significant role in technology, in the bioethanol production process valuable chemical properties and functions from lignin and hemicelluloses are not fully recovery, the black liquor result from process being using specially for energy recovery. About half of wood components are dissolved into this black liquor. The dissolved organic compounds consist mainly in degraded lignin and also hemicelluloses and cellulose degradation products. Also, phenols derived from biomass are valuable and useful chemicals, due to their pharmacological properties including antiviral inhibitor (anti-HIV). These compounds with good antioxidant activity can be used to preserve food from lipid peroxidation and oxidative damage occurring in living systems (Martínez et.al., 1996; Mahugo Santana et.al., 2009; Nenkova, et.al.2011). Antioxidants can also prevent the loss of food color, flavor and active vitamins content, providing the stabilization of the molecules involved in such characteristics. They can also be used for the production of adhesives and for the synthesis of polymer.

It is well known that, biomass also contains many other natural products: waxes and fatty acids, polyacetylenes, terpenoids (e.g., monoterpenoids, iridoids, sesquiterpenoids, diterpenoids, triterpenoids), steroids, essential oils, phenolics, flavonoids, tannins, anthocyanins, quinones, coumarins, lignans, alkaloids, and glycosidic derivatives (e.g., saponins, glycosides, flavonoid glycosides) (Alonso et.al., 1998; Japón-Luján et.al., 2006; Faustino, 2010; Fang et.al., 2009; Gallo, 2010; Carro, 1997; Kojima, 2004). In this regards, are needed more studies to recover these important compounds from biomass for use in pharmaceutical industry, food industry, and so.

2. Extraction techniques

Actually, there are known many different techniques used for biomass extraction: liquid-solid extraction, liquid-liquid extraction, partitioning, acid-base extractions, ultrasound extraction (UE), microwave assisted extraction (MAE). The capability of a number of extraction techniques have been investigated, such as solvent extraction (J.A. Saunders, D.E. Blume, 1981) and enzyme-assisted extraction (B.B. Li, B. Smith, M.M. Hossain, 2006). However, these extraction methods have drawbacks to some degree.

The choice of extraction procedure depends on the nature of the natural material and the components to be isolated. The main conventional extraction procedures are liquid-liquid extraction and liquid-solid extraction. For liquid-liquid extraction is using two different solvents, one of which is always water, (water-dichloromethane, water-hexane, and so). Some of the disadvantages of this method are: cost, toxicity and flammability (Kaufmann 2002; McCabe, 1956; Perry, 1988; Sarker et. al., 2006).

Solid-phase extraction (SPE) can be used to isolate analytes dissolved or suspended in a liquid mixture are separated from a wide variety of matrices according to their physical and chemical properties. Conventional methods include: soxhlet extraction, maceration, percolation, extraction under reflux and steam distillation, turbo-extraction (high speed mixing) and sonication. Although these techniques are widely used, have several shortcomings: are very often time-consuming and require relatively large quantities of polluting solvents, the influence of temperature which can lead to the degradation of thermo labile metabolites (Kaufmann 2002; McCabe, 1956; Sarker et. al., 2006; Routray, 2012).

Supercritical fluid extraction (SFE), microwave-assisted extraction (MAE) and pressurised solvent extraction (PSE) are fast and efficient unconventional extraction methods developed for extracting analytes from solid matrixes.

2.1. Supercritical fluid extraction

Supercritical fluid extraction (SFE) is one of the relatively new efficient separation method for the extraction of essential oils from different plant materials. The new products, extracts, can be used as

a good base for the production of pharmaceutical drugs and additives in the perfume, cosmetic, and food industries. Use of SFE under different conditions can allow selecting the extraction of different constituents. The main reason for the interest in SFE was the possibility of carrying out extractions at

temperature near to ambient, thus preventing the substance of interest from incurring in thermal denaturation.

Supercritical fluid extraction has proved effective in the separation of essential oils and its derivatives for use in the food, cosmetics, pharmaceutical and other related industries, producing high-quality essential oils with commercially more satisfactory compositions (lower monoterpenes) than obtained with conventional hydro-distillation (Ehlers et al., 2001; Diaz-Maroto et al., 2002; Ozer et al., 1996). Also, extraction with supercritical fluids requires higher investment but can be highly selective and more suitable for food products. This plays a mechanistic role in supercritical fluid chromatography (SFC), where it contributes to the separation of the solutes that are injected into the chromatographic system.

Supercritical fluid extraction is an interesting technique for the extraction of flavouring compounds from vegetable material. It can constitute an industrial alternative to solvent extraction and steam distillation processes (Stahl, E. and Gerard, D. 1985). SFE allows a continuous modification of solvent power and selectivity by changing the solvent density (Nykanen, I.et al., 1991). Nevertheless, the simple SFE process, consisting of supercritical CO_2 extraction and a one-stage subcritical separation, in many cases does not allow a selective extraction because of the simultaneous extraction of many unwanted compounds.

2.2. Ultrasound-assisted solvent extraction

Ultrasound assisted extraction is very efficient extraction procedure. Sonication induces cavitation, the process in which bubbles with a negative pressure are formed, grown, oscillated, and may split and implode. By this process different chemical compounds and particles can be removed from the matrix surface by the shock waves generated when the cavitation bubbles collapse. The implosion of the cavities creates microenvironments with high temperatures and pressures. Schock waves and powerful liquid micro jets generated by collapsing cavitation bubbles near or at the surface of the sample accelerate the extraction (R. Kellner et al., 2004). Ultrasonic assisted extraction has many advantages since it can be used for both liquid and solid samples, and for the extraction of either inorganic or organic compounds (S.L. Harper et al., 1983). If extracted from solid samples, different problems can occur: there is a possibility of the decomposition of the analyte which could be trapped inside of the collapsing cavitational bubbles. The ultrasound extraction system can be also applied as a dynamic system in which the analytes are removed as soon as they are transferred from the solid matrix to the solvent. In this process, furthermore, the sample is continuously exposed to the solvent (I. Rezic' et al., 2008).

This is a modified maceration method where the extraction is facilitated by the use of ultrasound (high-frequency pulses, 20 kHz). Ultrasound is used to induce a mechanical stress on the cells of biomass solid samples through the production of cavitations in the sample. The cellular breakdown increases the solubilization of metabolites in the solvent and improves extraction yields. The efficiency of the extraction depends on the instrument frequency, and length and temperature of sonication. Ultrasonification is rarely applied to large-scale extraction; it is mostly used for the initial extraction of a small amount of material. It is commonly applied to facilitate the extraction of intracellular metabolites from plant cell cultures (Kaufmann, 2002; Sarker, 2006).

2.3. Pressurized Solvent Extraction (PSE)

Pressurized solvent extraction or "accelerated solvent extraction," employs temperatures that are higher than those used in other methods of extraction, and requires high pressures to maintain the solvent in a liquid state at high temperatures. It is best suited for the rapid and reproducible initial extraction of a number of samples. The solid biomass sample is loaded into an extraction cell, which is placed in an oven. The solvent is then pumped from a reservoir to fill the cell, which is heated and pressurized at programmed levels for a set period of time. The cell is flushed with nitrogen gas, and the extract, which is automatically filtered, is collected in a flask. Fresh solvent is used to rinse the cell and to solubilize the remaining components. A final purge with nitrogen gas is performed to dry the material. High temperatures and pressures increase the penetration of solvent into the material and improve metabolite solubilization, enhancing extraction speed and yield. Moreover, with low solvent requirements, pressurized solvent extraction offers a more economical and environment-friendly alternative to conventional approaches

As the material is dried thoroughly after extraction, it is possible to perform repeated extractions with the same solvent or successive extractions with solvents of increasing polarity. An additional advantage is that the technique can be programmable, which will offer increased reproducibility. However, variable factors, e.g., the optimal extraction temperature, extraction time, and most suitable solvent, have to be determined for each sample (Kaufmann, 2002; Tsubaki, 2010; Sarker, 2006).

Microwave-assisted extraction (MAE) or simply microwave extraction is a relatively new extraction technique that combines microwave and traditional solvent extraction. The microwave energy has been investigated and widely applied in analytical chemistry to accelerate sample digestion, to extract analytes from matrices and in chemical reactions. Application of microwaves for heating the solvents and plant tissues in extraction process, which increases the kinetic of extraction, is called microwave-assisted extraction. Microwave energy is a non-ionizing radiation that causes molecular motion by migration of ions and rotation of dipoles, without changing the molecular structures if the temperature is not too high. Nonpolar solvents, such as hexane and toluene, are not affected by microwave energy and, therefore, it is necessary to add polar additives. Microwave-assisted extraction (MAE) is an efficient extraction technique for solid samples which is applicable to thermally stable compounds accepted as a potential and powerful alternative to conventional extraction techniques in the extraction of organic compounds from materials. The microwave-assisted extraction technique offers some advantages over conventional extraction methods.

Compared to conventional solvent extraction methods, the microwave-assisted extraction (MAE) technique offers advantages such as improved stability of products and marker compounds, increased purity of crude extracts, the possibility to use less toxic solvents, reduced processing costs, reduced energy and solvent consumption, increased recovery and purity of marker compounds, and very rapid extraction rates.

The use of MAE in natural products extraction started in the late 1980s, and through the technological developments, it has now become one of the popular and cost-effective extraction methods available today, and several advanced MAE instrumentations and methodologies have become available, e.g., pressurized microwave-assisted extraction (PMAE) and solvent-free microwave-assisted extraction (SFMAE).

Comparison between conventional and MAE extraction method

This technique has been used successfully for separation of phenolic compounds from types of biomass, polyphenols derivates, pyrimidine glycosides, alkaloids, terpenes, and so.

In most cases, the results obtained suggested that the microwave assisted method was more convenient even compared to the ultrasound extraction method.

Pyrimidine glycosides

The studies regarding the microwave extraction of vicine and convicine (toxic pyrimidine glycosides) from *Vicia faba* using a methanol: water mixture (1:1 v/v) involves two successive microwave irradiations (30 s each) with a cooling step in between. No degradation could be observed under these conditions,but further irradiation was found to decrease the yield of vicine and convicine. The yield obtained was 20% higher than with the conventional Soxhlet extraction method.

Alkaloids Sparteine, a lupine alkaloid, was extracted from *Lupinus mutabilis*, with methanol: acetic acid (99:1, v/v) in a common microwave oven and the microwave irradiation program used one to five cycles of 30 s with a cooling step in between and conduct to 20% more sparteine than was obtained with a shaken-flask extraction using the same solvent mixture for 20 min.

Terpenes Five terpenic compounds: linalool, terpineol, citronellol, nerol and geraniol, associated with grape (*Vitis vinifera*) aroma was extracted from must samples by MAE (Carro et al.,1997). Was investigated the influence of the parameters: extracting solvent volume, extraction temperature, and amount of sample and extraction time. Several conditions were fixed, such as the extraction time (10 min) and the applied power (475 W). The solvent volume appeared to be the only statistically significant factor, but was limited to 15 mL by the cell size. The highest extraction yield was obtained with both the solvent volume and the temperature at their maximum tested values. In contrast, the sample amount had to be minimized in order to obtain the best recoveries. The final optimized extraction conditions were as follows: 5 mL sample amounts extracted with 10 mL of dichloromethane at a temperature of 90°C for 10 min with the microwave power set at 50% (475 W).

Steroids Recently, was demonstrated that only 30–40 s were sufficient to extract ergosterol quantitatively by MAE using 2 mL methanol and 0.5 mL 2 M sodium hydroxide. Microwave irradiation was applied at 375W for 35 s and the samples were cooled for 15 min before neutralization with 1 M hydrochloric acid followed by pentane extraction. The yield was similar to or even higher than that obtained with the traditional methanolic extraction followed by alkaline saponification and pentane extraction.

Alkaloids The extraction of two alkaloids cocaine and benzoylecgonine by focused MAE was optimized by taking into account several parameters such as the nature of the extracting solvent, particle size distribution, sample moisture, applied microwave power and radiation time. MAE was found to generate similar extracts to those obtained by conventional SLE but in a more efficient manner. Indeed, 30s were sufficient to extract cocaine quantitatively from leaves, using methanol as solvent and a microwave power of 125 W. (Kaufmann, 2002).

Phenolic Compounds

In recent years, synthetic antioxidants were reported to have the adverse effects such as toxicity and carcinogenicity and this situation has forced scientists to search for new natural antioxidants from herbs or the other materials. Phenolic compounds, the most important bioactive compounds from plant sources, are among the most potent and therapeutically useful bioactive substances, providing health benefits associated with reduced risk of chronic and degenerative disease (Luthria, 2006; Tsubaki et al., 2010; Proestos, 2008).

Extraction is one of the most imperative steps in the evaluation of phenolic compounds from plant. Often is done a saponification prior to the extraction step because is necessary to cleave the ester linkage to the cell walls (Robbins, 2003).

The capability of a number of extraction techniques have been investigated, such as solvent extraction and enzyme-assisted extraction. However, these extraction methods have drawbacks to some degree. For example, solvent extraction is time consuming and enzyme in enzyme assisted extraction is easy to denature. In the case of Soxhlet extraction, the extraction time vary from 1 minute to 6 h. Ultrasonic is one of the most industrially used methods to enhance mass transfer phenomena (Japón-Luján et.al. 2006; Luthria, 2006; Pérez-Serradilla, 2007). Meanwhile, microwave assisted extraction heats the extracts quickly and significantly accelerates the extraction process (Martínez, 1996; Kojima, 2004; Patsias, 2009). Simultaneous ultrasonic/microwave assisted extraction (UMAE) coupled the advantage of microwave and ultrasonic, presenting many advantages (Kojima, 2004).

Extraction of phenolic compounds from solid samples is usually carried out by stirring (Luthria, 2006; Nepote, 2005), although the use of auxiliary energies has demonstrated to accelerate the process (Japón-Luján, et.al.2006; Pérez-Serradilla, 2007). Microwave-assisted extraction (MAE) is the process by which microwave energy is used to heat polar solvents in contact with solid samples and to partition compounds of interest between the sample and the solvent, reducing both extraction time and solvent consumption.

The conventional liquid–solid extraction techniques, such as heat reflux extraction (HRE), ultrasonic extraction (UE) and maceration extraction (ME), are discommodious, laborious, time-consuming and require large volumes of toxic organic solvents. So increasing attention is paid to the development of more efficient extraction methods for the rapid extraction of active compounds from materials.

The current analytical methods used to extract phenolic compounds from liquid samples are based on *liquid-liquid extraction* (LLE). Although this technique offers efficient and precise results, it is relatively time-consuming, possibly harmful due the use of large volume of organic solvents (frequently toxic) and highly expensive. For these reasons, there is an increasing tendency to replace LLE by solid-phase extraction (SPE) for liquid samples. SPE was developed in the 1980s, and has emerged as a powerful tool for chemical isolation and purification. This methodology is an alternative extraction to LLE due to it reduces organic solvents consumption, the length of analysis and it can be automated (Martínez, et. al., 1996; Kojima, 2004; Patsias, et.al., 2009).

Although most attention has been focused on the determination of phenolic compounds in aqueous samples, more substituted phenols, such as pentachlorophenol, show limited transport in water and they are more likely absorbed in sediments and soils. This fact contributes to the persistent of these compounds in the environment and it results in high concentrations of them that could affect aquatic and earth organism. For extraction, Soxhlet extraction is one of the most popular techniques for isolating phenolic compounds from solid samples, due to its simplicity, inexpensive extraction apparatus. Despite the good results obtained with this methodology, Soxhlet extraction makes the analysis procedure excessive time consuming. Moreover, it requires large amount of hazardous organic solvents.

Ultrasonic extraction is another conventional technique to extract analytes from solid samples. Although sonication is faster than Soxhlet extraction, it also requires large volumes of toxic and expensive organic solvents.

The studies show that the compounds are extracted more effectively when the energy provided by microwave is employed (Perez-Serradilla, 2011).

3. Experimental studies

The efficiencies of different solvents (water, acid and alcohol) in the extraction of caffeine and phenols from leaves of white, black, green and red tea in different solvents: ethanol, isopropanol, methanol and water. Extraction was performed comparative by ultrasonic and by MAE. Determination of the total amount of phenolic compounds was studied comparative using different extraction times 5, 15 and respectively 30 minutes. The microwave irradiation shortens time necessary to extract phenols and caffeine from tea samples (between 30 and 50 seconds). The results of the comparison investigation are presented in the figure 1.

4. Conclusion

Chromatographic determination of phenolic compounds isolated from the tea samples by ultrasonic and MAE extraction is comparable. The difference between the two methods of extraction consists in extraction time and amount of solvents used. Also, the yield for MAE was about 20% is 20% higher than that of the ultrasonic extraction.

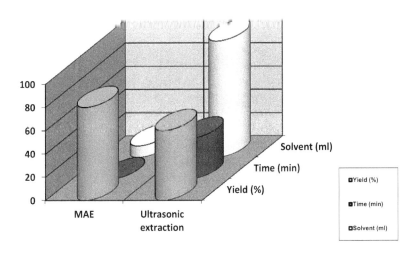

Figure 1. Comparison between the two extraction methods

Author details

Adina-Elena Segneanu, Paulina Vlazan,
Paula Sfirloaga and Ioan Grozescu
*National Institute of Research and Development for Electrochemistry and Condensed Matter –
INCEMC Timisoara, Romania*

Florentina Cziple
Eftimie Murgu University, Resita, Romania

Vasile Daniel Gherman
Politehnica University of Timisoara, Romania

5. References

Alonso M.C., Puig D., Silgoner I., Grasserbauer M., Barcelo´ D., (1998), Determination of priority phenolic compounds in soil samples by various extraction methods followed by liquid chromatography–atmospheric pressure chemical ionisation mass spectrometry, Journal of Chromatography A, 823, 231–239;

Carro,N., GarciaC.M., Cela R., (1997), Microwave-assisted extraction of monoterpenols in must samples, Analyst, 122 (4), 325–330;

Diaz-Maroto MC, Perez-Coello MS, Cabezudo MD. Supercritical carbon dioxide extraction of volatiles from spices – comparison with simultaneous distillation – extraction. J. of Chromatography A, 947, 23-29, 2002.

Ehlers D, Nguyen T, Quirin KW, Gerard D. Anaylsis of essential basil oils-CO₂ extracts and steam-distilled oils. Deutsche Lebensmittel-Rundschau, 97, 245-250, 2001.

Fang X., Wang J., Zhou H., Jiang X., Zhu L., Gao X.(2009), Microwave-assisted extraction with water for fast extraction and simultaneous RP-HPLC determination of phenolic acids in Radix Salviae Miltiorrhizae, J. Sep. Sci., 32, 2455 – 2461;

Faustino H., Gil N., Baptista C., Duarte A.P., (2010), Antioxidant Activity of Lignin Phenolic Compounds Extracted from Kraft and Sulphite Black Liquors, Molecules, 15, 9308-9322;

Gallo M., Ferracane R., Graziani G., Ritieni A. Fogliano V. (2010), Microwave Assisted Extraction of Phenolic Compounds from Four Different Spices, Molecules, 15, 6365-6374;

Harper S.LWalling., J.F., Holland D.M., Pranger L.J., Anal. Chem. 55 (1983) 1553.

Japón-Luján, R., Luque-Rodríguez, J.M., & Luque de Castro, M. D. (2006), Multivariate optimization of the microwave-assisted extraction of oleuropein and related biophenols from olive leaves, Analytical and Bioanalytical Chemistry, 385, 753–759;

Kaufmann B., Christen P., (2002), Recent extraction techniques for natural products: microwave-assisted extraction and pressurised solvent extraction, Phytochemical Analysis, Vol.13, 2, 105–113;

Kellner R., Mermet J., Otto M., Valcarel M., Widmer H.M., Analytical Chemistry, second ed., Wiley-VCH, Germany, 2004.

Kojima, M., Tsunoi, S., Tanaka, M., (2004), High performance solid-phase analytical derivatization of phenols for gas chromatography–mass spectrometry, J. Chromatogr. A 1042,1-7;

Li X.; Zeng, Z.; Zhou, J. (2004), High thermal-stable sol–gel-coated calix[4]arene fiber for solid-phase microextraction of chlorophenols, Anal. Chim. Acta 509, 27-37;

Li B.B., Smith B. Hossain M.M., Separation and Purification Technology 48 (2006) 189.

Luthria, D.L., Mukhopadhyay, S., Kwansa, A.L. (2006), A systematic approach for extraction of phenolic compounds using parsley (Petroselinum crispum) flakes as a model substrate, *Journal of the Science of Food and Agriculture*, 86, 1350–1358;

Mahugo Santana C., Sosa Ferrera Z., Torres Padrón M. E., Santana Rodríguez J.J., (2009), Methodologies for the Extraction of Phenolic Compounds from Environmental Samples: New Approaches, *Molecules, 14*, 298-320;

McCabe, W. L., and J. C. Smith, *Unit Operations of Chemical Engineering*, McGraw- Hill, 1956;

Martinson, L., Pasurull, L., Marec, R.M., Borrull, F., Calull M., (1996), Comparative study of the use of high-performance liquid chromatography and capillary electrophoresis for determination of phenolic compounds in water samples, *Chromatographia, 43*, 619-624;

Nenkova S., Radoykova T., Stanulov K., (2011), Preparation and antioxidant properties of biomass low molecular phenolic compounds (review), *Journal of the University of Chemical Technology and Metallurgy, 46, 2, 109-120*;

Nepote, V, Grosso, N.R., Guzman, C.A. (2005), Optimization of extraction of phenolic antioxidants from peanut skins, *Journal of the Science of Food and Agriculture*, 85, 33 38;

Nykanen, I., Nykanen, L. and Alkio, M. (1991) J. Essential Oil Res. 3, 229.

Ozer EO, Platin S, Akman U, Hortasçsu O. Supercritical Carbon Dioxide Extraction of Spearmint Oil from Mint-Plant Leaves. Can. J. Chem. Eng., 74, 920-928, 1996.

Patsias, J., Papadopoulou-Mourkidou, E. (2009), Development of an automated on-line solid-phase extraction–high-performance liquid chromatographic method for the analysis of aniline, phenol, *Molecules, 14*, 315;

Perry, R. H., and D. Green, *Perry's Chemical Engineers' Handbook*, 6th Edition, McGraw Hill, 1988;

Pérez-Serradilla, J.A., Japón-Luján, R., Luque de Castro, M.D. (2007), Simultaneous microwave-assisted solid–liquid extraction of polar and nonpolar compounds from Alperuja, *Analytica Chimica Acta*, 602, 82–88;

Proestos C., Komaitis M., (2008), Application of microwave-assisted extraction to the fast extraction of plant phenolic compounds, *LWT* 41, 652–659;

Rezic I., Krstic D., Bokic Lj., Ultrasonic extraction of resins from an historic textile, Ultrasonics Sonochemistry 15 (2008) 21 24

Robbins, R. J. (2003), Phenolic Acids in Foods: An Overview of Analytical Methodology, *J. Agric. Food Chem., 51*, 2866-2887;

Routray W., Orsat V., (2012), Microwave-Assisted Extraction of Flavonoids:A Review, *Food Bioprocess Technol*. 5, 409–424;

Sarker S.D., Latif Z., Gray A. I.,(2006), Natural Products Isolation Second Edition, *Humana Press* Inc.;

Saunders J.A., Blume D.E., Journal of Chromatography A 205 (1981) 147.

Stahl, E. and Gerard, D. (1985) Perfumer Flavorist 10, 29.

Sun,Y., Cheng, J. (2002), Hydrolysis of lignocellulosic materials for ethanol production: a review. *Bioresource Technology*, 83, 1–11;

Tsubaki S., Sakamoto M., Azuma J., (2010), Microwave-assisted extraction of phenolic compounds from tea residues under autohydrolytic conditions, *Food Chemistry*, 123,1255–1258;

Chapter 2

Lignocelluloses Feedstock Biorefinery as Petrorefinery Substitutes

Hongbin Cheng and Lei Wang

Additional information is available at the end of the chapter

1. Introduction

1.1. Lignocelluloses feedstock (LCF) biorefinery

1.1.1. Background

The material needs from our society are reaching the crisis point, as the demand for resources will soon exceed the capacity of the present fossil resource based infrastructure [1]. Currently, fossil-based energy resources, such as petroleum, coal, and natural gas, are responsible for about three quarters of the primary energy consumption in our world. While decreasing crude-oil reserves, enhanced demand for fuels worldwide, increased climate concerns about the use of fossil-based energy carriers, and political commitment, the focus has recently turned to develop the utilization of renewable energy resources [2]. Gullón et al. [3] described the variety of problems on present social, economic and technological situation, which including: the fear for a shortening of the supplies of basic resources, as the population growth; the increasing per capita demands of the developing economies for goods and energy, derived from the increasing purchase power of the population; environmental challenges, especially those related to effects of greenhouse gas emissions (emphasis on CO_2) on the global climate; the national security issues surrounding reliance on imported oil [4].

On our market, nowadays, there are more than 2500 different oil-based products. The petroleum crisis of the 1970s resulted in a shift from total reliance on fossil resources and simultaneously triggered research into biomass based technologies. As a result of the oil crisis, renewable resources became a popular phrase [5]. Currently, the most of energy requirements in the world are still met by fossil fuels. The limited deposits of these fossil fuels coupled with environmental problems have prompted people to look for sustainable resources as alternatives to meet the increasing energy demand. Bio-energy production has the advantage of forming smaller amounts of greenhouse gases compared to the conversion of fossil fuels, as the carbon dioxide generated during the energy conversion is consumed during subsequent biomass re-growth [6]. However, simply providing sustainable and non-polluting energy will not be enough. In our life, clothes, shelter, tools, medications and so on are all, to a greater or lesser degree, dependent on organic carbon. As fossil-based resources will be replaced, new sources of organic carbon have be found or alternate applications and processing of existing sources must be developed. The challenge is to find replacements not only for current usage, but also for the even future greater energy consumption, with a likely concomitant increase in biomass demand for manufacturing [7].

1.1.2. What is biorefinery?

The core aim for biorefineries is to produce both high-volume liquid fuels and high-value chemicals. As petroleum refinery uses petroleum as the major input and processes it into many different products, the term 'biorefinery' has been coined to describe the processing complexes that will use biomass as feedstocks to produce a wide spectrum of chemicals, fuels and bio-based materials, that can be used as industrial intermediates or sold directly to consumers [1, 8, 9]. Biorefineries have been considered as the key for access to an integrated production of chemicals, materials, goods, fuels and energy of the future [10]. As oil prices continue to rise and biorefining technology matures, biorefineries are playing an increasingly major role in the global economic system, with the potential to ultimately replace petroleum refineries as the world's principal method of fuel generation.

1.1.3. Lignocelluloses feedstock (LCF) biorefinery

The largest organic carbon reservoir in our world is the biomass - plants and algae. Each year, plants fix approximately 90 billion tons of CO_2, most of this as wood [11]. Lignocelluloses are the natural combination of cellulose, hemicelluloses and lignin. It's the raw material for potential conversion to energy fuels and chemical feedstock for manufacturing. LCF biorefinery has been defined as one of the so-called phase-III biorefinery concepts which are characterized by the ability to use a variety of resources by different routes to generate multiple products [12].

A LCF biorefinery uses lignocellulosic biomass, including forestry residue, agricultural residue, yard waste, wood products, animal wastes, etc. Initially, plant material is cleaned and broken down into the three main fractions (hemicellulose, cellulose, and lignin) by chemical digestion or enzymatic hydrolysis. Hemicellulose and cellulose can be produced by alkaline and acid. Lignin can also be further broken down with enzymes. The hemicellulose and cellulose are sugar polymers, which can be converted to their component sugars through hydrolysis. A hemicellulose is a polymer that contains five-carbon sugars (usually D-xylose and L-arabinose), six-carbon sugars (D-galactose, D-glucose, and D-mannose), and uronic acid. Cellulose is a polymer of only glucose. The hydrolysis process of hemicelluloses and cellulose result in the aforementioned sugars [13].

The LCF Biorefinery is a promising alternative due to the abundance and variety of available raw materials and the good position of the conversion products on the market [14]. Its profitability is also dependent on the technology employed to alter the structure of lignocellulosic biomass in order to produce high value co-products from its three main fractions *i.e.* cellulose, hemicellulose, and lignin [15].

Currently the main feedstock for biorefineries is still based on starch. The practiced technologies in fuel ethanol industry are primarily based on the fermentation of sugars derived from starch and sugar crops, which are quite mature with little possibility of process improvements. However, using starch and sugar crops to produce ethanol also has been questioned since it draws its feedstock from a food stream. Lignocellulosic biomass is a more promising renewable resource as it is available in large quantities and does not compete with food or feed. Lignocellulosic biomass is a renewable resource that stores energy from sunlight in its chemical bonds, with great potentials for the production of affordable fuel ethanol [16, 17]. Its main obstacle for a major breakthrough is the high production costs for bioenergy products.

On the other hand, lignocellulosic biomass derived products can significantly reduce green house gas emissions, compared to fossil-based products. Also, many common petrochemicals could be obtained with lower green house gas emissions from bio-based feedstocks. The maturity and economics of the conversion processes and logistics is a major challenge for lignocellulosic biomass [18].

1.1.4. The main goal of Biorefinery

With, implementing innovative, environmentally sound and cost-effective production technologies for a variety of products, the integrated biorefinery is increasing the availability and use of bioenergy and bio-based products. The main objective of a biorefinery is to produce high-value low volume and low-value high-volume products by a series of producing processes. The processes are designed to maximize the valued products while minimizing the waste streams by converting low-value high-volume intermediates into energy. The high-value products can enhance the profitability, and the high-volume fuels will help to meet the global energy demand. The power produced from a biorefinery can also help to reduce the overall cost. Figure 1 shows the elements of a biorefinery, in which biomass is used to produce various useful products such as fuel, power, and chemicals by biological and chemical conversion processes [13].

Traditionally, the matured biorefinery pathways include bioconversion (aerobic and anaerobic digestion) and chemical conversion (bio-pulping). There are two most promising emerging biorefinery platforms. One is the sugar platform and the other is the thermo-chemical platform (syngas platform). In sugar biorefineries platform, biomass will be broken down into different types of component sugars for fermentation or other biological processing into various fuels and chemicals. In thermo-chemical biorefineries platform, biomass will be synthesized hydrogen and carbon monoxide or pyrolysis oil, the various components of which could be directly used as fuel [19].

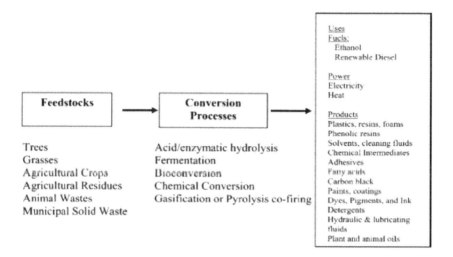

Figure 1. Simple procedure for three-step biomass-process-products [13]

1.1.5. Disadvantages

It is very important to increase the reaction rates, as slow reactions rates is one of the main disadvantages for biological conversions in biorefinery processes. Another disadvantage is the often low product concentrations, which means the high product recovery costs with existing technology. The lower yields of targeted products is often found in some multiple products systems [20]. Therefore the biorefinery processes to become an actual alternative to fossil fuels and petroleum-derived products, biorefinery processes must be competitive and cost-effective [21].

2. Biofuels

As an important category of bioenergy, biofuel is a type of fuel which is biologically derived from biomass. The biofuels, which include liquid, solid biofuels and various biogases, can replace the conventional petroleum or petroleum derived products. Many biological reactions involved in biofuels production are at mild conditions, can offer relatively high products yields and generally result in low levels of contamination to the environment. The modern application of biological transformations, known as biotechnology, is also an evolving field that has great promise for substantial improvements and significant cost reductions. In this section, several liquid and gases biofuels are introduced e.g. (1) bioethanol, biobutanol, and biodiesel which can replace the gasoline used as transportation fuels; and (2) biogas, which is produced from anaerobic digestion of biomass as a substitute for natural gas either for industrial applications or for transportation.

2.1. Bioethanol

Bioethanol is a promising transport fuel alternative to gasoline because it has higher oxygen content and no sulphur or nitrogen when compared with gasoline [22]. Currently, the blends E5 and E10 that consist of 5% (v/v) and 10% (v/v) ethanol respectively, have a widespread usage since these blends can supply the existing vehicular fleet without major changes to engines. High bioethanol blends (E100, E95 and E85) require modified or dedicated vehicles.

Bioethanol can be produced from three types of raw materials: sugars (from sugarcane, sugar beet, molasses, and fruits), starch (from corn, cassava, potatoes, and root crops) and cellulose (from wood, agricultural residues, waste sulphite liquor from pulp and paper mills). Among the three main types of raw materials, cellulose contained in lignocellulosic biomass represents the most abundant global source of biomass, which can be utilised for bioethanol production [23]. There are also two approaches for producing bioethanol from lignocellulosic biomass through (1) Biochemical (2) Thermochemical processes.

2.1.1. Biochemical production of bioethanol

Figure 2 illustrates the high level technologies for producing bioethanol from these various biomass feedstocks. Typically, the common steps for biologically producing bioethanol from different feedstocks are fermentation and distillation. For the first generation (1G) bioethanol production, the sugar extracted from sugar-rich crops and that from starch digestion by amylases or acids is directly fermented to bioethanol. To convert lignocellulosic biomass into second generation (2G) bioethanol, an additional step of pre-treatment is usually required.

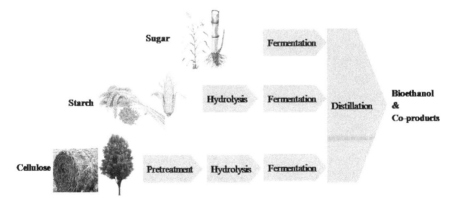

Figure 2. Technologies required producing bioethanol from biomass. [24]

A wide variety of lignocellulosic feedstocks are potentially available for bioethanol production such as wood, grass, agricultural waste and MSW (municipal solid waste). Their physical structures and chemical compositions are different; therefore technologies applied for bioethanol production can be diverse. In addition to the main product bioethanol, co-products are also usually produced, such as heat and electricity generated by burning lignin-rich residue from fermentation and also, potentially, a wide range of high value-added chemicals like acetic acid, furfural and hemicellulose sugar syrup and the low molecular weight lignin.

General technologies required for biologically producing 2G bioethanol include (1) pre-treatment, (2) enzymatic hydrolysis, (3) fermentation, and (4) distillation.

Pre-treatment is applied to enhance the accessibility of enzyme to biomass by increasing available biomass particle surface area for enzyme to attack. This can be achieved by partially removing lignin and/or hemicellulose, changing the structure of biomass fibres to decrease cellulose crystallinity and its degree of polymerization. The current available pre-treatment methods can be classified as mechanical, chemical and biological. Table1 summarised some typical pre-treatment methods and their characterisations. Pre-treatment has been viewed as the most expensive step in the biologically production of bioethanol. Therefore, it is important to assess the economic feasibility of the pre-treatment method in addition to its technology performance. More information about each pre-treatment method can be found in Section 5.

Enzymatic hydrolysis is carried out under mild conditions with potentially high sugar yields and relatively low maintenance costs. Nevertheless, major challenges for cost-effective commercialisation remain, such as the high cost of enzymes, the slow rate of enzymatic reaction and potential inhibition by sugar degradation products from pre-treatments [48]. In enzymatic hydrolysis, cellulose is hydrolysed by a suite of enzymes, including cellulase and β-glucosidase crudely purified from lignocellulose-degrading fungi such as Trichoderma reesi, Trichoderma viride and Aspergillus niger. Cellulase refers to a class of enzymes including endocellulase breaking internal bonds of cellulose, exocellulase cleaving from the free ends of chains produced by endocellulase to form cellobiose (a dimer of glucose), and cellobiase (β-glucosidase) then hydrolysing cellobiose to produce glucose monomers. In addition, most of cellulase mixtures contain hemicellulase that facilitates hemicellulose hydrolysis to assist with the overall effectiveness of enzymatic hydrolysis.

After the enzymatic hydrolysis, sugar monomers can then be fermented to ethanol by micro-organisms (e.g. Saccharomyces cerevisiae and Zymomonas mobilis). Fermentation has been commercialised in brewery and food manufacturing for centuries and itself is not a complex and expensive process. The challenges regarding fermentation for the bioethanol industry are: (1) to convert pentose (C5 sugar) which cannot be fermented by the conventional yeast efficiently, and (2) to prevent inhibition caused by sugar degradation products from pre-treatments. Research has shown the feasibility of construction and application of genetically engineered yeasts capable of converting both pentose and hexose to ethanol [49]. Further potential lies in using bacteria with the metabolic pathways necessary to ferment all sugars available from lignocellulosic biomass. Z. mobilis has shown to be capable of metabolising 95% of glucose, 80% of xylose and 40% of other sugars in corn stover hydrolysate [50]. Metabolic engineered Geobacillus thermoglucosidasius has demonstrated an ethanol yield of over 90% of theoretical at temperatures in excess of 60°C [51].

Pre-treatment method	Process and conditions	Possible changes in biomass	Disadvantages	Reference
Steam explosion	No agent temperature:160-260°C,20-50 bar , 2-5 minutes	Dissolve hemicelluloses Low sugar degradation	Partially degrade hemicellulose	[25-27]
Ammonia fibre explosion (AFEX)	Ammonia as agent 65-90°C, 0.5-3 hours	Change biomass physical structure Increasing hemicelluloses hydrolysis	Limited effective on soft and hardwood	[28, 29]
SO₂/H₂SO₄ explosion	SO₂ as agent, 160-220°C, < 2 minutes	Dissolve hemicelluloses effectively for hardwood and agricultural residues	Degradation of hemicelluloses, less effective for softwood	[30, 31]
CO₂ explosion	CO₂ as agent, 35°C, 56.2 bar, 10-60 minutes	Interrupt crystalline structure of cellulose	Inefficient for softwood and high capital cost	[32, 33]
Hot liquid water	Water as agent, 190-230°C, 45 seconds-4 minutes	Effectively dissolve hemicelluloses Very low degradation	Water recycling prohibitively expensive	[34-36]
Dilute acid	H₂SO₄ as agent , over 160°C, 2-10 minutes	Effectively dissolve hemicelluloses	Needs neutralisation, significant formation of fermentation inhibitors	[37-39]
Alkaline	NaOH/ Ca(OH)₂ /Ammonia as agent, 70-120°C, 20-60 minutes	Removal of lignin Low hemicelluloses degradation	Costs of reagents and wastewater treatment are high	[40-42]
Oxidation	Ca(OH)₂+O₂/H₂O₂ as agent, 140°C, 3 hours	Removal of lignin Low hemicelluloses degradation	Costs of reagents and wastewater treatment are high	[43, 44]
Organic solvent	Ethanol as agent, 140-200°C, 30-150 minutes	Removal of lignin	Cost of solvent recovery is high	[45, 46]
Ionic liquid	Ionic liquid as agent, 120°C, 22 hours	Remove of lignin and hemicellulose	Costs of reagents and long treatment time	[47]

Table 1. Chemical pre-treatment methods for lignocellulosic biomass.

Bioconversion process configurations, including Separate Hydrolysis and Fermentation (SHF), Simultaneously Saccharification and Fermentation (SSF), Simultaneously Saccharification and Co-Fermentation (SSCF), and Consolidated Bioprocessing (CBP). The SHF has many advantages, such as allowing both enzyme and micro-organisms to operate at their optimum conditions. Also, any accidental failure of enzymatic hydrolysis and fermentation would not affect the other steps. Alternatively the enzymatic hydrolysis may also be combined with fermentation and can thus be carried out simultaneously in a same reactor - this being known as the simultaneous saccharification and fermentation (SSF). During enzymatic hydrolysis, the cellulases are strongly inhibited by hydrolysis products: glucose and short cellulose chains ('end-point' inhibition). SSF can overcome this inhibition by fermenting the glucose to ethanol as soon as it appears in solution. However, ethanol itself inhibits the action of fermenting micro-organisms and cellulase although ethanol accumulation is less inhibitory than high concentrations of hydrolysis products [52]. Nevertheless, SSF operating at the compromised temperature (37-40 °C) has some drawbacks caused by the different optimal temperatures for the action of cellulases (45-50° C) and the growth of microorganisms (typically 28-35°C). One method to overcome this disadvantage is the utilisation of thermo-tolerant fermenting organisms. SSCF is a promising SSF process where the micro-organism co-ferment pentose and hexose to bioethanol. CBP currently becomes the focus of most research efforts to date; it integrates cellulase production, cellulose hydrolysis and fermentation in one step by using an engineered strain [53]. Many studies have been reported in CBP technologies developments recently [54-56].

Nevertheless, other significant efforts are also required to enable future integrated biorefinery. They include (1) promising process designs to integrate energy consumption and minimise the water footprint (2) producing a range of high value added by products, *e.g.* power, chemicals, and lignin-derived products *etc.*

2.1.2. Thermo-chemical production of bioethanol

The thermo-chemical bioethanol production refers to a series of processes including biomass indirect gasification, alcohol synthesis and alcohol separation as shown in Figure 3.

The biomass is processed and dried by flue gas before being fed to biomass gasifier. The biomass is chemically converted to a mixture of syngas components (*i.e.* CO, CH_4, CO and H_2 *etc*), tars, and a solid char which is the fixed carbon residual from the biomass. The heat required for endothermic gasification reactions is supplied by circulating hot synthetic olivine 'sand' between the gasifier and combustor. The solid char and 'sand' from the gasifier are separated by cyclones and then sent to a char combustor where the char is oxidised by oxygen injected. The heat released from the oxidation of the char reheats the 'sand' over 980 °C. The hot 'sand' is then sent to the gasifier to provide heat required by gasification reactions. The ash from the char combustor and sand particles captured are sent to landfill after being cooled and moistened. The tar produced in the gasifier is reformed to CO and H_2 with the presence of catalyst in a bubbling fluidized bed reactor. The syngas generated in the biomass gasifier goes through a cooling and clean-up process to remove CO_2 and H_2S. During this process, the tar is reformed in an isothermal fluidized bed reactor and the catalyst is regenerated. The cleaned syngas is then converted to alcohols in a fixed bed reactor. The produced alcohol stream is depressurised in preparation of dehydration and separation afterwards. The evolved syngas in alcohol stream is recycled to the Gas Cleanup & Conditioning section. Finally, the alcohol mix is separated to methanol, ethanol and other higher molecular weight alcohols. The heat required for the gasifier and reformer operations and electricity for internal power requirements is provided by a conventional steam cycle. The steam cycle produces steam by recovering heat from the hot process streams throughout the plant.

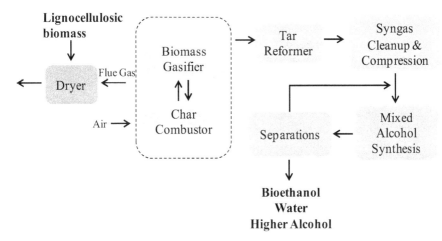

Figure 3. The schematic of a thermo-chemical cellulosic ethanol production process [57]

To compare these two approaches (biochemical *vs.* thermo-chemical) for producing bioethanol from economic point of view, process simulation and economic analysis are usually performed to calculate the minimum ethanol selling price (MESP) calculated from the discounted cash flow method. The MESP is defined as the selling price of bioethanol that makes the net present value of the biomass to bioethanol process equal to zero with a certain discounted cash flow rate with in a return period over the life of the plant [37]. In other words, it refers to the ethanol price at the break-even point which means annual costs and income are equal at this price. Several studies suggested that the estimated prices for 2G bioethanol produced biochemically is in the range of 2.16 to 4.44 USD $/gallon, depending on the type of biomass feedstock, technologies applied and the reference year based on [37, 58-61]. On the other hand, NREL (National Renewable Energy Laboratory) reported a relatively low MESP for bioethanol produced thermo-chemically as 1.07 USD $/gallon. Nevertheless, raw materials cost (including biomass feedstock and catalyst or enzyme) is the main contributor to the MESP. For example, the cost of corn stover accounts for 40% and 43% of the MESP for bioethanol biochemically and thermo-chemically produced respectively [37, 57].

From environmentally point of view, a comparative LCA study showed that biochemical approach offers a slightly better performance on greenhouse gas emission and fossil fuel consumption impact categories, but the thermo-chemical pathway has significantly less water consumption [62].

2.2. Butanol

Butanol is another attention attracted alternative fuel to gasoline besides ethanol because of its properties with respect to gasoline blending, distribution and refuelling, and end use in existing vehicles. For instance, butanol has relatively high energy content which is 30% higher than ethanol and is closer to gasoline. Additionally, butanol has low vapor pressure, low sensitivity to water and it is less volatile, and less flammable when compared with other liquid fuels [63]. Therefore butanol can be handled conventionally in the existing petroleum infrastructure, including transport *via* pipeline. It also can be blended, at any ratio, with either gasoline or diesel fuel at existing refineries, thus avoiding the capital investment associated with plant revamps and the need for major operational, *etc.*

Similarly to bioethanol, butanol can be biochemically produced from both agricultural crops and lignocellulosic biomass using *Clostridium acetobutylicum* or *C. beijerinckii* to ferment lignocellulosic hydrolysate sugars (hexoses and pentoses) to butanol. Traditionally, sugar-rich agricultural crops such as corn, cane molasses and whey permeate have been successfully used as feedstocks in the commercial production of butanol for decades. However, the cost for these food crops rises significantly nowadays; therefore, lignocellulosic biomass becomes more popular as substrates for butanol production. In similar ways of producing bioethanol, pre-treatments are required prior to enzymatic hydrolysis (using cellulase and cellobiose). However, one of technology challenges is the inhibition caused by by-products in pre-treatments such as furfural, HMF, acetic acid, and ferulic acid generated in dilute acid pre-treatments *etc.* Among these by-products, ferulic and o coumaric acids were found can significantly inhibit fermentation but furfural and HMW were surprisingly stimulating to the cell culture [64].

The resulted lignocellulosic hydrolysate is then fermented by microorganisms *via* Acetone-butanol-ethanol (ABE) fermentation (Figure 4). The main challenge in the ABE fermentation is the product butanol itself is toxic to the fermenting microorganisms. In order to overcome this drawback, focused research efforts are to (1) improve the fermentation strategies to minimise the level of inhibitors accumulated such as simultaneously removing butanol and (2) to develop or genetically improve butanol – producing cultures.

However, biobutanol has several potential shortcomings. It is more toxic to humans and animals in the short term than ethanol or gasoline (although some components of gasoline, such as benzene, are more toxic and/or carcinogenic). And it is not clear whether butanol will degrade the materials commonly used in automobiles that can come into contact with motor fuels; building evidence suggests that it will not cause problems, but there has been no definitive testing on the wide range of potentially affected polymers and metals [65].

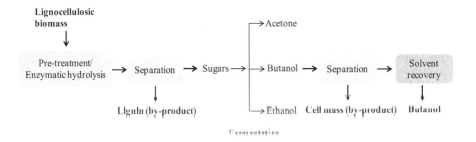

Figure 4. Phases of ABE fermentation for producing butanol

Additionally, butanol is reported cannot deliver a better economic feasibility and a more sustainable environmental performance when compared with bioethanol under the current level of technology [66]. The relatively low yield of solvents out of glucose (mixture of acetone, ethanol and butanol), which is in the range of 33% - 45% (wt), is the main cause for the high cost of butanol. This economic study argued that butanol perhaps can be sold as chemicals rather than transport fuel unless the technology would be improved to make butanol production economically competitive with bioethanol.

2.3. Biodiesel

Biodiesel refers to a liquid fuel alternative to petroleum diesel which can be used alone or blended with petroleum diesel. Similarly to bioethanol blends, blends of 20% biodiesel (B20) or lower can be used in diesel equipment without or with only minor modifications. Biodiesel can be produced from animal fat or oil from plants such as soybean and *Jatropha*, or from microalgae and fungi.

2.3.1. Biodiesel from vegetable oil

Conventionally, the biodiesel is produced from vegetable oil with the presence of alcohol/alkaline/acid catalyst. This process is known as transesterificaiton or alcoholysis as shown in Figure 5 [67].

The vegetable oil is converted to esters and glycerol by reacting with an alcohol which can be ethanol, methanol or butanol. During this reaction, catalysts (*e.g.* alkalis, acids or enzymes) are required to improve the reaction rate and yield. Alkalis including NaOH, KOH and carbonates *etc.* are usually used as catalyst when feedstock containing less than 4% fatty acids. Acids, which are normally used when feedstocks contain more than 4% free fatty acids, include sulfuric acid, hydrochloric acid and sulphonic acids *etc.* Lipase, an enzyme that catalyses the hydrolysis of fats, can be used as a biocatalyst [68].

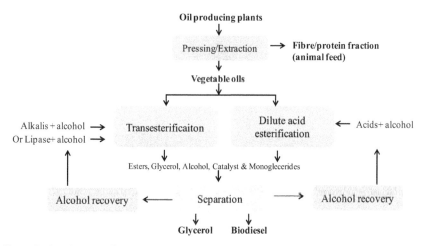

Figure 5. The schematic of biodiesel production [67]

A review by Ma and Hanna [69] summarized the parameters significantly influencing the rate of transesterificaiton reaction which include temperature, ratio of alcohol to oil, type of catalyst and catalyst concentration. The ester yield is increased by rising the transesterificaiton temperature; however, it will increase the risk of forming methanol bubbles when the temperature is close to methanol's boiling point. The ratio of alcohol to oil depends on the type of catalyst used which is approximately 6:1 for alkali catalyst and 30:1 for acid catalyst [70]. Enzyme used as a catalyst is becoming more attractive nowadays because it tolerates free fatty acid and water contents in the oil to avoid soap formation and thus results in an easier purification of biodiesel and glycerol [68]. However, the relatively high price of enzyme catalyst makes its utilization in the commercial production of biodiesel challenging.

Nowadays, 90% of U.S. biodiesel is made from soybean oil. The price relationship between vegetable oils and petroleum diesel is key influential factor to the profitability of biodiesel industry. Because of the increasing price of vegetable oils, biodiesel industry is suffering uncomfortable situations [71]. As a result, alternative non-food feedstocks and the associated technologies are becoming the focused research in biodiesel area.

Jatropha curcas is an agro-forestry crop growing in tropical and sub-tropical countries, such as India, Sahara Africa, South East Asia and China. This crop grows rapidly and takes 2-3 years to reach maturity with economic yields [72]. Lu *et al.* reported a higher than 98% biodiesel yield by a pre-esterification using solid acid followed by a transesterificaiton using KOH [73]. A high yield of 98% (wt) is also reported by Shah *et al.* [74] which is obtained from *Jatropha* oil using *Pseudomonas cepacia* lipase. Kumari *et al.* [75] also documented a relatively high yield of 94% (wt) biodiesel yield from *Jatropha* oil using lipase from *Enterobacter aerogene*. They also reported negligible loss in lipase activity even after repeated use for several cycles.

2.3.2. Biodiesel from microalgae

Due to biodiesel produced from oil crops, waste cooking oil and animal fat cannot meet the high demand for renewable transport fuels, another biomass feedstock microalgae becomes attractive. This is because (1) microalgae are sunlight-driven cells, (2) grow rapidly with biomass double time of 24 hours, (3) require less high quality land used compared to other feedstock, (4) many are exceedingly rich in oil and (5) biodiesel produced from microalgae is 'carbon neutral' [76] (see Figure 6). However, several challenges need to be tackled in order to produce biodiesel from microalgae commercially. Stull et al. [77] provides a comprehensive review discussing these challenges and potential tackles.

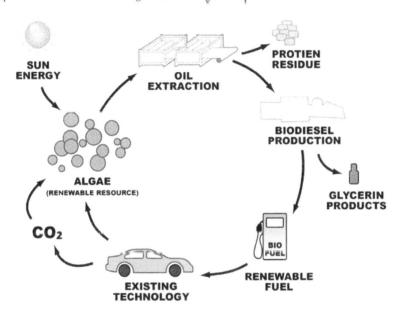

Figure 6. Life cycle of biodiesel produced from microalgae

There are estimated 300 000 species in algal strain. After screening, typical species including *Botryococcus braunii*, *Nannochloropsis sp.*, *Neochloris oleoabundane*, *Nitzschia sp.*, and *Schizochytrium sp.* have up to 77% (dry wt) oil content [76]. Microalgal biomass is produced with the presence of light, fed carbon dioxide and essential inorganic elements including nitrogen (N), phosphorus (P), iron and in some cases silicon. Biomass is then harvested and extracted to obtain oil for biodiesel production using transesterificaiton with methanol. Nutrients and spent biomass are recycled in the downstream process.

Factors involved in these phases are all important to be considered and optimized to maximize the biomass yield and minimize the production cost. First of all, the light level needs to be manipulated to deliver an optimal light to all of the algae cells within the culture. The excess light level not only can results in less efficient use of absorbed light energy but also can cause biochemical damage to the photosynthetic machinery [77]. Secondly, though minimal nutrients requirement can be estimated according to the approximate molecular formula of microalgae which is $CO_{0.48}$ $H_{1.83}$ $N_{0.11}$ $P_{0.01}$[78], nutrients such as phosphorous must be supplied in excess. In order to minimise the nutrient cost, sea water supplement with commercial nitrate and phosphate fertilisers can be used for growing microalgae [76]. Thirdly, the choice of facility (open raceway ponds or closed photobioreactor) is important since the scale-up of biomass production is largely depending on the surface area rather than volume because light only penetrate a few centimeters [77]. The former raceway pond is an open-top close loop recirculation channel with a typical depth of 0.3 m. It is relatively cheap to build and has been operated with extensive experience for decades. However, the drawbacks for this type of facility are (1) it is difficult to avoid microbial contamination, (2) it requires for extensive areas of land for ht raceways and substantial cost regarding harvesting, and (3) it has poorly mixed therefore has optically dark zone [76, 77]. The photobioreactor a tubular reactors consists of an array of glass or plastic transparent tubes. It requires a large amount of energy for pumping and compressing air for sparging culture [77].

The biomass broth from production phase is harvested and processed to remove water and residual nutrients which are recycled. The concentrated biomass paste is then extracted to obtain oil and lipids using water and extraction solvent (*e.g.* hexane) [79]. It is difficult to release lipids from microalgae intracellular location using an energy-efficient way because of the large amount of solvent required. Also it is key to avoid significant contamination by other cellular components such as DNA [77].

The efforts in academic research and industrial commercialization of biodiesel production from microalgae include: (1) integration of production process such as energy integration, water and nutrient recycling; (2) improvement of microalgae biology *via* genetic and metabolic engineering such as enhancing their photosynthetic efficiency, increasing biomass yield and oil content and improving temperature tolerance to reduce cost associated with cooling; (3) improving photobioreactors regarding their capacity and operational ability [76].

2.4. Biogas from anaerobic digestion

Anaerobic digestion (AD) has been used to treat biodegradable solid waste such as MSW, industrial waste and sewage sludge over decades. Biogas containing methane and carbon dioxide is the main product form AD digester. Generally, biogas is collected in the gas tank and they can be directly exported to national gas grid or sent to combustion in the CHP system to generate electricity (with a yield in the range of 0.7 – 2.0 kwh/m³biogas) and heat.

AD process is a dynamic complex system involving microbiological, biochemical, and physical-chemical processes though which the biodegradable waste are turned into biogas. Among biological waste treatment methods, AD has been identified as the most environmentally sustainable option for treating biowaste since it offers a unique technology which enables not only diverting biodegradable from landfill but also producing bio-energy and potential by-products such as a beneficial soil conditioner [80].

AD systems generally have four classifications [80]:

- Mesophilic (30 - 40 °C) or thermophilic (50 - 65°C) according to temperature
- Wet digestion (< 15% dry solid) or dry digestion (between 20% - 40% dry solid) according to the solid content in feedstock
- Single step (one vessel) or multiple step digestions (normally two-step digestion i.e. hydrolysis and methanogenesis)
- Batch digestion (loading feedstock in the beginning and remove products at the end of process) or continuous digestion (loading feedstock and withdraw products continuously)

Generally, five microbial groups are considered to be important to the process such as hydrolysing bacteria, acidogenic bacteria, acetogenic bacteria, aceticlastic and hydrogentrophic methanogens. They are involved in a series of digestion steps which are described as following and in Figure 7 [81] :

1. Carbohydrates, lipids, proteins *etc.* are broken down through hydrolysis to sugars, long – chain fatty acids and amino acids by extracellular enzymes released by hydrolytic bacteria;
2. Then these molecules are converted into volatile fatty acids, alcohols, CO_2 and H_2 in acidogenesis step;
3. These molecules are then further converted by acetogenic bacteria mainly into acetic acid, H_2 and CO_2;
4. Finally, all these intermediate products are turned into CH_4, CO_2 and water in the last step where methanogenic bacteria are involved. Three biochemical pathways are used by methanogens to produce methane gas:

   a. Acetotrophic methanogenesis: $4\ CH_3COOH \rightarrow 4\ CO_2 + 4\ CH_4$

   b. Hydrogenotrophic methanogenesis: $CO_2 + 4\ H_2 \rightarrow CH_4 + 2\ H_2O$

   c. Methylotrophic methanogenesis: $4\ CH_3OH + 6\ H_2 \rightarrow 3\ CH_4 + 2\ H_2O$

Due to for different substances, biological consortia and digestion conditions, the overall biogas yield and methane content will vary. Typically, the methane content of biogas is in the range of 40-70 % (v/v) [82].

Several key factors influence the Ad performance. They include pH, temperature, organic loading rate (OLR), the ratio of inoculum to substance (I/S) and the presence of inhibitory substances. Generally, mesophilic AD (35 - 37 °C) is more preferred than thermophilic AD (50 - 60°C) since the latter one offers less methane yield and it is more sensitive to environment change [81]. The pH range suggested for AD process is in the range of 6.8 -7.2 [80]. In addition, anaerobic digestion requires attention to the loading of nutrients for bacteria including carbon and nitrogen. The proper ratio of these two components (C/N) depends on the digestibility of the carbon and nitrogen sources between 20: 1 and 30:1. Other nutrients such as S, Mg, K, P, Ca, Fe, Zn, Al, Ni, Co, Cu and vitamin B12 are necessary [80]. However, these components are generally contained in the Organic Fraction of MSW (OFMSW) while they are added in the laboratory scale AD systems.

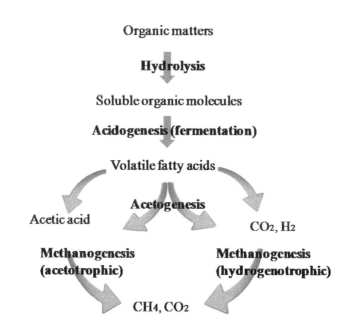

Figure 7. Anaerobic digestion biochemical conversion pathways

Regarding the AD process operation, I/S ratio is considered as one the most important parameter. It is suggested to be approx 1 by Raposo et al.[83] who found that biogas production was inversely proportional to the I/S ratio in the range of 1 to 3. two stage AD is more preferred because it provides optimum environmental conditions for each bacteria group, offers accelerated digestion rates, better stability and thus increased methane yield [80]. Another process parameter is retention time which includes hydraulic retention time and solid retention time. The former refers to the mean time that any portion of liquid feed remains in a digestion system; the latter is defined as the mean time for any portion of solid feed or microbal biomass remains in the digester. These two retention times are the same in a single stage digester; while in a two stage digestion system the longer solid retention time is, the higher degradation rates and biogas yields are obtained [80]. In addition to the above process parameters discussed, the organics loading rate (OLR) is also critical which is measured as volatile solid (VS) or chemical oxygen demand (COD) of feed to a unit volume of digester per unit time [80]. The range for OLR is suggested in the range of 6 - 9.7 kg VS/day/m³ which is varied with the biodegradability of feedstock and AD systems [81].

Furthermore, the quality of OFMSW treated in biogas plants is also crucial for balanced performance of the biogas process, the technical feasibility of process and the use of residual/effluent as agricultural soil conditioner. Therefore, the costs associated with waste collection, sorting and pre-treatment should be considered [84].

Currently, most of MSW in the U.S. are sent to composting as an alternative to landfill. This is because it is more difficult to treat OFMSW than treating wastewater or manure. In addition, the AD of OFMSW requires a large amount of investment and technological experience as well as a higher capital and operating cost than composting and landfilling [82]. The relatively low gate fees for landfill in the U.S. and relatively low energy prices make AD difficult to be commercialized in the U.S compared to those in Europe [82]. However, in the UK, there is currently very little waste treatment using AD apart from the use of AD from the co-digestion of manure and wastewater [85].

However, LCA studies have shown that AD of MSW reduces environmental impacts and in more cost - effective (in Europe) on a whole system basis than composting or landfilling options [86, 87].

3. Commodity chemicals and materials

Today, only a small numbers of chemicals are produced from lignocellulosic feedstocks via fermentation. Much less attention has been given to biomass as a feedstock for organic chemicals, while there has been a strong political and technical focus on using biomass to produce transportation fuels. However, replacement of petroleum-derived chemicals with those from biomass will play a key role in sustaining the growth of the chemical industry [88]. One way to replace petroleum is through biological conversion of lignocellulosic resources into products now derived from petroleum. The current developments especially in fermentation technologies, membrane technologies and genetic manipulation open new possibilities for the biotechnological production of market relevant chemicals from renewable resources [5].

In lignocellulosic feedstocks biorefinery processes, the sugars or some of the fermentation products can be chemically converted into a variety of chemicals, which could be used to form biological materials, such as protein polymers, xantham gum, and polyhydroxybutyrate. The lignin as remaining fraction from lignocellulosic feedstocks, could be converted through hydrogenation processes into materials, such as phenols, aromatics, and olefins, or simply burned as a boiler fuel for cost efficiency of the overall process. Currently, conventional chemicals include acetaldehyde, acetic acid, acetone, n-butanol, ethylene, and isopropanol can simply be derived from LCF. Appropriate organisms could then convert the sugars into the desirable products and co-products for this process. The advantage to such products is that the market is already established, and minimal effort is required to integrate these products into existing markets. However, co-product markets might be limited, and caution must be taken in considering their impact on overall economics, especially for large-scale implementation. A sequence of processes comparable to those employed for cellulosic ethanol production would be used to pre-treat the lignocellulosic biomass to open its structure for the weight of the feedstock. Therefore, lignocellulosic biomass might be expected as the low cost of raw materials could be converted to a variety of commodity chemicals and materials [20].

3.1. Present promising commodity chemicals and materials from LCF biorefinery

3.1.1. Lactic acid

Lactic acid represents a chemical with a small world market, and the market for traditional applications of lactic acid is estimated to be growing at about 3–5% annually. New products based on lactic acid may increase the world market share significantly, which includes the use of derivatives such as ethyl esters to replace hazardous solvents like chlorinated hydrocarbon solvents in certain industrial applications. In theory, one mole of glucose results in almost two moles of lactic acid. The recovery process for lactic acid is much more sophisticated than that of the ethanol fermentations, involving various precipitations, chromatographic and distillation steps [5].

Lactic acid can be converted to methyl lactate, lactide, and polylactic acid (PLA) by fermentation [89]. The PLA is a biodegradable polymer used as environmentally friendly biodegradable plastic, which can be the replacement for polyethylene terephthalates (PETs) [90]. Recently, attempts have been made to produce PLA homopolymer and its copolymer by direct fermentation by metabolically engineered [91], shows a great potential for utilizing lignocellulosic feedstock for the key biodegradable polymers. Efforts are also under way to develop efficient processes for converting biologically produced lactic and hydroxypropionic acids to methacrylic and acrylic acids [88].

Lactic acid can be produced either chemically or by microbial fermentation. A major disadvantage for chemical synthesis is the racemic mixture of lactic acid. Microbial fermentation offers both utilization of renewable carbohydrates and production of pure L- or D-lactic acid depending on the strain selected. Currently, most of lactic acid production is produced mainly from corn starch. However, the use of lignocellulosic feedstock for lactic acid production appears to be more attractive because they do not impact the food chain for humans. But the process for converting lignocellulosic feedstock into lactic acid is not cost efficient due to the high cost of cellulase enzymes involved in cellulose hydrolysis [92, 93]. In addition, the main bottleneck during the hydrolysis of lignocellulosic feedstock by cellulases is the inhibition on cellulase by glucose and cellobiose, which remarkably slows down the rate of lignocellulosic feedstock hydrolysis [94]. Economic improvements on the process are mainly focused on increasing the lactic acid tolerance, reducing the requirements for complex and cost intensive growth supplements and products recovery [95].

3.1.2. Acetone–butanol–ethanol (ABE)

An acetone – butanol – ethanol blend (in a ratio of 3-6-1) may serve as an excellent car fuel, which can be easily mixed not only with petrol but also with diesel. ABE as a fuel additive has the advantage of a similar heat of combustion to hydrocarbons, and perfect miscibility with hydrocarbons, even when water is present. The fermentative production of ABE used to be the second largest industrial fermentation after ethanol production [5]. Product inhibition caused principally by butanol is the main problem that hindering commercial development of the fermentation process. One way to overcome this inhibition problem would be to couple the fermentation process to a continuous product removal technique, so that inhibitory product concentrations are never reached. However, even with continuous product removal, product formation in these systems does not proceed indefinitely, because of the inhibition caused by the accumulation of mineral salts in the reactor [96]. Due to the shortage of raw materials, namely corn and molasses, and to decreasing prices of oil, ABE fermentation is not profitable when compared to the production of these solvents from petroleum. During the 1950s and 1960s, ABE fermentation was replaced by petroleum chemical plants.

Currently, the production of mixtures of acetone, butanol and ethanol (ABE) by sugars derived from lignocellulosic feedstocks continues to receive attention because of its potential commercial significance. The traditional fermentative production of acetone–butanol– ethanol is batch anaerobic bacteria fermentation with Clostridia. The substrate consists of molasses, and phosphate and nitrogen sources. Instead of molasses other sugar sources like sugar from lignocellulosic feedstock can also serve as a raw material for fermentation [97].

3.2. Xylan

As one of main polysaccharides in lignocellulosic biomass, xylan has a variety of applications in our everyday life and affects our well-being. For example, (1) xylans are important functional ingredients in baked products [98]; (2) xylans can be potentially used for producing hydrogels as biodegradable coatings and also encapsulation matrices in many industrial applications; (3) xyl, the main constituent from xylans, can be converted to xylitol which is used as a natural food sweetener and a sugar substitute [99]; (4) xylans can be used for clarification of juices and improvement in the consistency of beer [100]; (5) xylans are also important for livestock industry as they are critical factors for silage digestibility; (6) xylans are major constituents in non-nutritional animal feed [101]; (7) xylans can be converted to sugars and then further to fuels and chemicals; (8) enzymes that degrade xylan can facilitate paper pulping and biobleaching of pulp [100].

Xylans, the main component in hemicellulose, are heteropolysaccharides with homopolymeric backbone chains of 1,4 linked β-d-xylopyranose units. In addition to xylose, xylans may also contain arabinose, glucuronic acid or its 4-O- methyl ether, acetic, ferulic, and p-coumaric acids. Xylans can be categorized as linear homoxylan, arabinoxylan, glucuronoxylan, and glucuronoarabinoxylan. Depends on the different sources of xylan (i.e. soft- and hard- wood, grasses, and cereals), the composition of xylans differs [100].

Hemicellulose can be derived via chemical treatment or enzymatic hydrolysis. As discussed in Section 2.1.1, several pre-treatments listed in Table 1 are available to fractionate, solubilize and hydrolyze and separate hemicellulose from cellulose and lignin components. Generally, hemicelluloses are solublized by either high temperature and short residence time (270°C, 1 min) or lower temperature and longer residence time (190 °C, 10 min) [102]. However, some of chemical treatment result in hemicellulose degradation by-products such as furfural and

5-hydroxymethyl furfural (HMF) which are inhibitors for microorganisms involved in downstream fermentation if applicable.

Biodegradation of xylan requires enzymes including endo-β-1,4-xylanase, β-xylosidase, and several accessory enzymes, such as α-L-arabinofuranosidase, α-glucuronidase, acetylxylan esterase, ferulic acid esterase, and p coumaric acid esterase, which are necessary for hydrolyzing various substituted xylans. The endo-xylanase attacks the main chains of xylans while β-xylosidase breaks xylooligosaccharides to monomeric sugar xylose. The α-arabinofuranosidase and α-glucuronidase remove the arabinose and 4-O-methyl glucuronic acid substituents from the xylan backbone [100]. The esterases hydrolyze the ester linkages between xylose units of the xylan and acetic acid (acetylxylan esterase) or between arabinose side chain residues and phenolic acids , for example ferulic acid (ferulic acid esterase) and p-coumaric acid (p-coumaric acid esterase) [100].

Hemicellulose hydrolysates from lignocellulosic biomass either obtained by chemical treatment or enzymatic hydrolysis are attractive feedstock for producing bioethanol, 2,3-butanediol or xylitol. Other value added products from hemicellulose hydrolysate include (1) ferulic acid, and (2) lactic acid which can be used in the food, pharmaceutical, and cosmetic industries [100].

3.3. Other main chemicals and materials from lignocellulosic feedstock

Acetic acid, at present, most demand of the commercial acetic acid is met synthetically. The production involves fermentation by a species of Acetobacter, which converts ethanol to acetic acid with a small final concentrations percentage (4–6%), using almost exclusively for vinegar production. In commercial practice, the actual yield roughly 75–80% of the theoretical yield [5].

Ferulic acid, as a precursor for numerous aromatic chemicals used in the chemistry industry, can be produced from lignocellulosic feedstock [88].

Levulinic Acid, Formic Acid and Furfural, their biorefinery process usually involves the use of dilute acid as a catalyst but it differs from other dilute acid lignocellulosic fractionating processes in that free monomer sugars are not the product. Instead, these monosaccharides are converted into the platform chemicals levulinic acid and furfural as the final products by multiple acid-catalysed reactions [103].

3.4. Opportunities and challenges

New products from lignocellulosic feedstock including new adhesives, biodegradable plastics, degradable surfactants, and various plastics and polymers could also be derived through the unique biotechnologies. The products with desirable properties that are not easily matched by petrochemical processing are particularly promising targets. Therefore, less price pressure would exist initially for such new products. However, to have a substantial impact on petroleum consumption, it is necessary to ensure that large markets have to be eventually resulted [20].

Even today, the potential of microorganisms for the production of bulk chemicals is far from being fully exploited. The cost of feedstocks still remains one of the crucial points if biotechnological processes are to succeed. The transition of industrial chemical production from petrochemical to biomass feedstock faces real hurdles. Biorefinery processes do not require the high pressures and temperatures compared with most non-biological chemical processes, thus have the potential to reduce costs. However, current non-biological chemical processes (often continuous, and well integrated) for production of commodity chemicals have become highly efficient by evolved through considerable investment. Therefore biorefinery processes for production of commodity chemicals must rapidly approach similar levels of efficiency and productivity. Nevertheless, available technologies, economic opportunities, and environmental imperatives make the use of lignocellulosic feedstock and biorefinery for industrial chemical production not only feasible but highly attractive from multiple perspectives [88].

Simple criteria have been devised to allow rapid screening of potential chemicals and materials from lignocellulosic feedstock for their economic merit. We now need to identify products that have economic potential and improve the technology to a point where these technologies can be applied in a cost-effective way [20].

4. Fractionation of lignocellulosic feedstock

4.1. Definition

Conversion of lignocellulosic materials to higher value products requires fractionation of the material into its components: lignin, cellulose, and hemicellulose, which convert to fuels, and chemicals for the production of most of our synthetic plastics, fibres, and rubbers is technically feasible. Liquefaction of LCF might serve as feedstocks for cracking to chemicals in the similar way that crude oil is presently used. Currently commercial products of LCF fractionation include levulinic acid, xylitol, and alcohols [104]. The ultimate goal of LCF fractionation is the efficient conversion of lignocellulose materials into multiple streams that contain value-added compounds in concentrations that make purification, utilization, and/or recovery economically feasible [15].

Fractionation of LCF is being developed as a means to improve the overall biomass utilization. Hemicellulose when separated from the LCF may find broader use for chemicals, fuel, and food application. The lignin separated in the process can be used as a fuel [105]. Unlike the lignin generated from pulping process, lignin fractionated from biomass by our approach is relatively clean, free of sulphur or sodium.

Fractionation of lignocellulosic materials is very difficult to accomplish efficiently, because of their complex composition and structure [106, 107]. However, fractionation of lignocellulosic materials is essential for some important applications, for example, paper-making, and in their conversion into basic chemical feedstocks or liquid fuels.

Figure 8 shows that fractionation of lignocellulosic biomass into its three major components, cellulose, hemicelluloses and lignin. It has been proposed as the first step of LCF refining to high value-added products [108]. Achieving high fractionation yields and maintaining the integrity of the macromolecular fractionation products are of major importance regarding the effectiveness of the whole refining process [109].

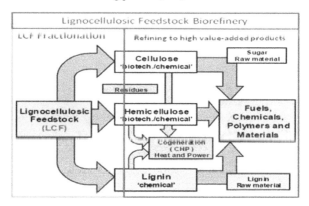

Figure 8. Lignocellulosic Feedstock Biorefinery [110]

4.2. Organosolv fractionation

The organosolv process is a unique and promising LCF fractionation. Using organosolv, lignocellulosic biomass can be converted into cellulosic fibres, hemicellulose sugars and low molecular weight lignin fractions in one-step fractionation [111-113]. Organosolv fractionation is the process to using organic solvents or their aqueous solutions to remove or decompose the network of lignin from lignocellulosic feedstocks with varying simultaneous hemicellulose solubilisation [114]. In this process, an organic or aqueous organic solvent mixture with or without an acid or alkali catalysts is used to dissolve the lignin and part of the hemicellulose, leaving reactive cellulose in the solid phase [106, 115-117]. Usually, the presence of catalyst can increase the solubilisation of hemicellulose and the digestibility of substrate is also further enhanced [110]. Comparing to other chemical pre treatments the main advantage of organosolv process is that relatively pure, low molecular weight lignin is recovered as a by-product [119]. Organic solvents are always easy to recover by distillation and recycled for fractionation; the chemical recovery in organosolv fractionation processes can separate lignin as a solid material and carbohydrates as syrup, both of which can be used as chemical feedstocks [112, 120, 121]. A variety of organic solvents have been used in the organosolv process such as ethanol, methanol, acetone, ethylene glycol, triethylene glycol, tetrahydrofurfuryl alcohol, glycerol, aqueous phenol, aqueous n-butanol, esters, ketones, organic acids, *etc* [117, 119, 122]. For economic reasons, among all possible solvents, the use of low-molecular-weight alcohols with lower boiling points such as ethanol and methanol has been favoured [123].

Organic solvents are costly and their use requires high-pressure equipment due to their high volatility. The applied solvents should be separated from the system is necessary because the residual solvents may be inhibitors to enzymatic hydrolysis and fermentation [106], and they should be recycled to reduce operational costs. Otherwise organic solvents are always expensive, so it should be recovered as much as possible, but this causes increase of energy consumption.

The organosolv fractionation seems more feasible for biorefinery of lignocellulosic biomass, as it considers the utilization of all the biomass components. However, there are inherent drawbacks to the organosolv fractionation. In order to avoid the re-precipitation of dissolved lignin, the fractionated solids have to be washed with organic solvent previous water washing, the cumbersome washing processes means more cost. In addition, organosolv fractionation must be performed under extremely tight and efficient control due to the volatility of organic solvents. No digester leaks can be tolerated because of inherent fire and explosion hazard [121]. Its successful commercialization will depend on the development of high-value co-products from lignin and hemicelluloses [124].

4.3. Ionic liquids fractionation

The ionic liquids (ILs) is a group of promising green solvents for the efficient fractionation of lignocellulosic materials. This technology has been used for delignification of lignocellulosic materials in paper-making [125]. Moreover, by fractionating lignocelluloses with ionic liquids it is possible to extract cellulose cleanly, which establishes a platform for the development of cellulose composites and derivatives.

ILs are liquid salts exist at relatively low temperatures (often at room temperature), which typically composed of large organic cations and small inorganic anions. By adjusting the anion and the alkyl constituents of the cation, ILs' solvent properties can be varied. The solvent properties include chemical and thermal stability, non-flammability, low vapour pressures and a tendency to remain liquid in a wide range of temperatures [126]. ILs are called "green" solvents, as no toxic or explosive gases are formed.

Most ILs are nonflammable and recyclable solvents with very low volatility and high thermal stability. Carbohydrates and lignin can be simultaneously dissolved in ILs, and the intricate network of non-covalent interactions between biomass polymers of cellulose, hemicellulose, and lignin is effectively disrupted while minimizing formation of degradation products [127-129].

ILs can dissolve large amounts of cellulose at considerable mild conditions and feasibility of recovering nearly 100% of the used ILs to their initial purity makes them attractive [130]. ILs as cellulose solvents, comparing with regular volatile organic solvents of biodegradability, possesses several advantages including low toxicity, broad selection of anion and cation combinations, low hydrophobicity, low viscosity, enhanced electrochemical stability, thermal stability, high reaction rates, low volatility with potentially minimal environmental impact, and non-flammable property.

However, ILs fractionation using ionic liquids faces many challenges in putting these potential applications into industrial scale., for example, the high cost of ILs, regeneration requirement [16]. Their toxicity toward enzymes and microorganisms must also be established before ILs can be considered as a real option for LCF pre-treatment [129].

Other main challenges are the recovery of ionic liquids and the recovery of hemi-cellulose and lignin from the ionic liquids after extraction of cellulose [126].

4.4. Liquid hot water (LHW) fractionation

Liquid hot water fractionation does not employ any catalyst or chemicals. Pressure is utilized to maintain water in the liquid state at elevated temperatures (160–240 °C) and provoke alterations in the structure of the lignocelluloses [131-133]. LCF in LHW undergoes high temperature cooking in water with high pressure. LHW pre-treatment has been reported to have the potential to enhance cellulose digestibility, sugar extraction, and pentose recovery, with the advantage of producing hydrolysates containing little or no inhibitor of sugar fermentation [134].

Water is an abundant, non-toxic, environmentally benign and inexpensive solvent. LHW is the part range of sub-critical water that near its critical point (374 °C, 22.1 MPa), Sub-critical water (SCW) possesses marvellous properties which are very different from that of ambient liquid water [135-138]. In SCW, dielectric constant, surface tension, and viscosity decrease dramatically with increasing temperature, which enhances the solubility of organic compounds. Sub-critical water is more like non-polar organic solvent (similar with acetone), thus it can substitute for some of organic solvents, and become a clean medium for chemical reactions. SCW is a tunable reaction medium for conducting ionic/free radical reactions, and an effective medium for energy and mass transfer. The ionic product of SCW is larger by three orders of magnitude than that of ambient water, which means concentrations of hydrogen and hydroxide ions are much higher. Therefore, in addition to the increase in kinetic rates with temperature, both acid and base catalyses by water are enhanced in SCW, which can be a solvent or reactant participated in chemical reaction. And without any pollution, hydrolysis in SCW is an environment-friendly technology

The objective of the liquid hot water is to solubilise mainly the hemicellulose to make the cellulose more accessible and to avoid the formation of inhibitors. By keeping the pH between 4 and 7 the autocatalytic formation of fermentation inhibitors are avoided during the fractionation [34, 139, 140]. If catalytic degradation of sugars occurs it results in a series of reactions that are difficult to control and result in undesirable side products.

The slurry generated after pre-treatment can be filtered to obtain two fractions: one solid cellulose-enriched fraction and a liquid fraction rich in hemicellulose derived sugars [34]. Lignin is partially depolymerised and solubilised as well during hot water fractionation but complete delignification is not possible using hot water alone, because of the re-condensation of soluble components originating from lignin.

Water under high pressure can penetrate into the LCF, hydrate cellulose, and remove hemicellulose and part of lignin. The major advantages are no addition of chemicals and no requirement of corrosion-resistant materials for hydrolysis reactors in this process. Liquid hot water pre-treatments are attractive from no catalyst requirement and low-corrosion potential. Liquid hot water has the major advantage that the solubilised hemicellulose and lignin products are present in lower concentrations, due to higher water input and subsequently concentration of degradation products like furfural and the condensation and precipitation of lignin compounds is reduced. However, water demanding in the process and energetic requirement are higher and it is not developed at commercial scale [141]

4.5. Combined technology for LCF fractionation

The efficiency of lignocelluloses utilization can be significantly improved by fractionation [40]. Fractionation of lignocellulosic materials may be achieved by various physical, chemical and biological methods. Combination of different methods may lead to more efficient fractionation processes of lignocellulosic materials [5].

The most promising combined technology for LCF fractionation is the combination of liquid hot water (LHW) with the assisted technologies, which usually are performed before or during the LHW fractionation, including steam explosion, CO2 explosion, Ammonia fibre explosion (AFEX), acid or alkaline pre-treatment, High energy radiation pre-treatment, Wet oxidation and Ozonolysis *etc*.

4.5.1. Combination with steam explosion

Steam explosion is the most widely employed physical-chemical pre-treatment for lignocellulosic biomass. It is a hydrothermal pre-treatment in which the biomass is subjected to pressurised steam for a period of time ranging from seconds to several minutes, and then the pressure is suddenly reduced and makes the materials undergo an explosive decompression. The treatment leads to the disruption of the structure of the material due to the rapid expansion of the water vaporized inside it. The temperatures involved are higher than, or close to, the glass transition temperature of hemicellulose, lignin and cellulose impregnated with water [142, 143], so that the internal cohesion of lignocelluloses is weakened and disaggregation and defibration of the material are facilitated. This pre-treatment combines mechanical forces and chemical effects due to the hydrolysis (auto-hydrolysis) of acetyl groups present in hemicelluloses.

Hydrolytic treatments of lignocellulosic biomass by saturated steam, with (un-catalyzed) and without (catalyzed) addition of small amounts of mineral acids, have been widely studied as a method to weaken the lignocellulosic structure and increase its chemical reactivity and enzyme accessibility [144, 145].

Un-catalyzed steam-explosion is one of only a very limited number of cost-effective pre-treatment technologies that have been advanced to pilot scale demonstration and commercialized application [16]. Autohydrolysis takes place when high temperatures promote the formation of acetic acid from acetyl groups; furthermore, water can also act as an acid at high temperatures. The mechanical effects are caused because the pressure is suddenly reduced and fibres are separated owing to the explosive decompression. In combination with the partial hemicellulose hydrolysis and solubilisation, the lignin is redistributed and to some extent removed from the material [146]. Catalyzed steam-explosion is very similar to un-catalyzed steam-explosion on their action modes, except that some acidic chemicals (gases and liquids), primarily including SO_2, H_2SO_4, CO_2, oxalic acid, etc. are used as catalysts to impregnate the LCF prior to steam-explosion, to improve recovering both cellulose and hemicellulose fractions [147]. It is recognized as one of the most cost-effective pre-treatment processes [148, 149]. Compared to un-catalyzed steam explosion, catalyzed steam-explosion has more complete hemicellulose removal leading to more increased enzymatic digestibility of LCF with less generation of inhibitory compounds [150]. A steam-explosion/separation process offers several attractive features when compared to the alternative hydrolysis and pulping processes. These include the potential for significantly lower environmental impact, lower capital investment, more potential for energy efficiency, less hazardous process chemicals and conditions [151]. Steam-explosion allows the recovery of all constitutive LCF components without the destructive degradation of any one component in favour of any other [152]. The process is generally followed by fractionation steps in order to separate the various components.

4.5.2. Combination with CO_2 explosion

Carbon dioxide explosion can also be used for lignocellulosic biomass pre-treatment. The method is based on the utilization of CO_2 as a supercritical fluid, which refers to a fluid that is in a gaseous form but is compressed at temperatures above its critical point to a liquid like density. Supercritical carbon dioxide has been used as an extraction solvent for non-extractive purposes, due to some advantages such as availability at relatively low cost, non-toxicity, non- flammability, easy recovery after extraction, and environmental friendly [153]. Besides a liquid-like solvating power, supercritical carbon dioxide displays gas-like mass transfer properties [154]

Supercritical pre-treatment conditions can effectively increase substrate digestibility by removing lignin. Addition of co-solvents such ethanol can improve delignification. Supercritical carbon dioxide has been mostly used as an extraction solvent but it is being considered for non-extractive purposes due to its many advantages [155]. CO_2 molecules are comparable in size to water and ammonia and they can penetrate in the same way the small pores of lignocelluloses. This mechanism is facilitated by high pressure. After CO2 explosive, pressure released, disruption of cellulose and hemicellulose structure is observed and consequently accessible surface area of the substrate to enzymatic attack increases [141].

4.5.3. Combination with ammonia fibre explosion (AFEX)

Similar to steam explosion, AFEX is one of the alkaline physical-chemical pre-treatment processes. Here the biomass is exposed to liquid ammonia under high pressure for a period time, and then the pressure is suddenly released, resulting in a rapid expansion of the ammonia gas that causes swelling and physical disruption of LCF fibres and partial decrystallization of cellulose. This swift reduction of pressure opens up the structure of lignocellulosic biomass leading to increased digestibility of biomass.

One of the main advantages of AFEX pre-treatment is no formation of some types of inhibitory by-products, which are produced during the other pre-treatment methods, such as furans in steam explosion pre-treatment.

AFEX has been studied for decreasing cellulose crystallinity and disrupt lignin-carbohydrates linkages [156]. Ammonia recovery and recycle is feasible despite of its high volatility [157] but the associated complexity and costs of ammonia recovery may be significant regarding industrial scale using of the AFEX pre-treatment [34, 158].

There are some disadvantages in using the AFEX process compared to some other processes. AFEX simultaneously de-lignify and solubilize some hemicellulose while decrystallizing cellulose, but does not significantly solubilize hemicellulose as acid and acid-catalyzed steam-explosion pre-treatments [159-161]. The AFEX produces only a pre-treated solid fraction, while steam explosion produces a slurry that can be separated in a solid and a liquid fractions [15]. Furthermore, ammonia must be recycled after the pre-treatment to reduce the cost and protect the environment [106, 158].

4.5.4. Combination with acid or alkaline treatment

A way to improve the effect of LHW fractionation is to add an external acid or alkali, which can catalyze the solubilisation of the hemicellulose, reduce the optimal pre-treatment temperature and gives a better enzymatic hydrolysable substrate [162-164].

Acid pre-treatments can be performed with concentrated or diluted acid. However utilization of concentrated acid is less attractive for ethanol production due to the formation of inhibiting compounds, and high acid concentration (e.g. 30-70%) in the concentrated-acid process makes it extremely corrosive and dangerous [165, 166]. Diluted acid pre-treatment appears as more favourable method for industrial applications and have been studied for fractionation wide range of lignocellulosic feedstocks, including softwood, hardwood, herbaceous crops, agricultural residues, wastepaper, and municipal solid waste. It performed well on most biomass materials, mainly xylan, but also converting solubilised hemicellulose to fermentable sugars. Of all acid-based pre-treatment methods, sulphuric acid has been most extensively studied since it is inexpensive and effective. Organic acids such as fumaric or maleic acids are appearing as alternatives to pre-treat LCF for fractionation. Organic acids also can pre-treat lignocellulosic materials with high efficiency although fumaric acid was less effective than maleic acid. Furthermore, less amount of furfural was formed in the maleic and fumaric acid pre-treatments than with sulphuric acid [167]. Phosphoric acid, hydrochloric acid and nitric acid have also been tested [34].

Alkali pre-treatment refers to remove lignin and a part of the hemicellulose, by use of alkaline solutions such as NaOH and Ca(OH)2, and efficiently increase the accessibility of enzyme to the cellulose. Alkali pre-treatment can be used at room temperature and times ranging from seconds to days. It is reported to cause less sugar degradation than acid pre-treatment. It is basically a delignification process, in which a significant amount of hemicellulose is solubilised as well. Alkaline pre-treatment of lignocellulosic materials causes swelling, increasing the internal surface of cellulose and decreasing the degree of polymerization and crystallinity, which provokes lignin structure disruption, and separation of structural linkages between lignin and carbohydrates [117]. In general, alkaline pre-treatment is more effective on hardwood, herbaceous crops, and agricultural residues with low lignin content than on softwood with high lignin content [168]. Alkali pre-treatment was shown to be more effective on agricultural residues than on wood materials [169]. Addition of an oxidant agent (oxygen/H_2O_2) to alkaline pre-treatment (NaOH/Ca(OH)$_2$) can favour lignin removal to improve the performance [170].

4.5.5. Combination with ammonia and carbon dioxide solution

The aim of combination is to enhance alkaline or acidic intensity of liquid hot water by ammonia or carbon dioxide for lignocelluloses fractionation.

Ammonia is an extremely important widely used bulk chemical. The polarity of Ammonia molecules and their ability to form hydrogen bonds explains to some extent the high solubility of ammonia in water. In aqueous solution, ammonia acts as a base, acquiring hydrogen ions from H2O to yield ammonium and hydroxide ions.

$$NH_3(aq) + H_2O(l) \approx NH_4{^+}(aq) + OH{^-}(aq)$$

The production of hydroxide ions when ammonia dissolved in water gives aqueous solutions of ammonia the characteristics of alkaline properties.

Carbon dioxide can be considered as an ideal solvent for the treatment of natural products, because of the relatively low critical pressure (73.8 atm) and critical temperature (31.1 °C), it. In contrast with organic solvent, Super-critical carbon dioxide is non-toxic, non-flammable, non corrosive, cheap and readily available in large quantities with high purity [171].

Carbon dioxide dissolves in water becomes acidic due to the formation and dissociation of carbonic acid:

$$CO_2 + H_2O \approx H_2CO_3 \approx H^+ + HCO_3{^-}$$

Over the temperature range 25-70 °C and pressure range 70-200 atm, the pH of solution ranged between 2.80 and 2.95, and increases with increasing temperature and decreases with increasing pressure [172]. It was shown that in the presence of water, supercritical CO_2 can efficiently improve the enzymatic digestibility of lignocellulosic materials [32].

4.5.6. Combination with high energy radiation treatment

Digestibility of lignocellulosic materials can be enhanced by the application of high energy radiation methods, such as microwave heating [173-175] and ultrasound [176, 177]. The treatments can cause hydrolysis of hemicellulose, and partial depolymerization of lignin, the increase of specific surface area, decrease of the degrees of polymerization and crystallinity of cellulose.

Microwave treatment is a physical-chemical process involving both thermal and non-thermal effects. Treatments can be carried out by immersing the biomass in dilute chemical reagents and exposing the slurry to microwave radiation for a period of time [178]. The treatment of ultrasound on lignocellulosic biomass have been used for extracting hemicelluloses, cellulose and lignin [179]. Some researchers have also shown that saccharification of cellulose is enhanced efficiently by ultrasound pre-treatment [100]. The efficiency of ultrasound in the treatment of vegetal materials has been already proved [101]. The well known benefits from ultrasounds, such as swelling of vegetal cells and fragmentation due to the cavitational effect associated to the ultrasonic treatment. Furthermore, mechanical impacts produced by the collapse of cavitation bubbles, give an important benefit of opening up the solid substrates surface for enzymatic hydrolysis [180].

However, the high energy radiation methods are usually energy-intensive and prohibitively expensive; appear to be strongly substrate-specific. The current estimation of overall cost from high energy radiation techniques looks too high, lack commercial appeal.

4.5.7. Combination with oxidative treatment

Wet oxidation

Wet oxidation is an oxidative pre-treatment method which employs oxygen or air as catalyst, and can be operated at relatively low temperatures and short reaction times [182]. It is an exothermic process, therefore self-supporting with respect to heat while the reaction is started [183]. Wet oxidation of the hemicellulose fraction is a balance between solubilization and degradation. Wet oxidation has been proven to be an efficient method for separating the cellulosic fraction from lignin and hemicellulose [184], and also been widely used for ethanol production followed by SSF [185]. Wet oxidation pre-treatment mainly causes the formation of acids from hydrolytic processes, as well as oxidative reactions. The hemicelluloses are extensively cleaved to monomer sugars, cellulose is partly degraded, and the lignins undergo both cleavage and oxidation in wet oxidation pre-treatment. Therefore lignin produced by wet oxidation cannot be used as a fuel [186]. In general, low formation of inhibitors and efficient removal of lignin can be achieved with wet oxidation pre-treatment.

Ozonolysis

Ozone is a powerful oxidant that shows high delignification efficiency [106]. This method can effectively degrade lignin and part of hemicellulose. The pre-treatment is usually carried out at room temperature, and does not lead to inhibitory compounds [187]. It is usually performed at room temperature and normal pressure and does not lead to the formation of inhibitory compounds that can affect the subsequent hydrolysis and fermentation. However, ozonolysis might be expensive since a large amount of ozone is required, which can make the process economically unviable [106].

5. Other bioconversion technologies

5.1. Landfill gas (LFG) production

As discussed in Section 2.4, anaerobic digester is a suitable waste treatment method to deal with wastewater, sewage sludge and animal mature since the high solid content of other types waste would challenge the anaerobic digester operation technologies. Currently most of biodegradable waste is sent to landfill where landfill gas (LFG) is generated.

Because the wastes sent to landfill include not only biodegradable components but also other hazard wastes, the LFG produced contains approx 40 - 60% methane, CO_2, and varying amounts of nitrogen, oxygen, water vapour, volatile organics (VOC), H_2S and other contaminates (also known as non-methane organic compounds NMOCs). Some other inorganic contaminants, for example, heavy metals are found present in the LFG. Therefore, the direct release of the landfill gas to atmosphere will cause serious greenhouse gas emissions and pollutions. LFG produced from landfill site has to be monitored and managed appropriately. The general LFG managing options are: flaring (burn without energy recovery), boiler (produces heat), internal combustion (producing electricity), gas turbine (producing electricity), fuel cell (producing electricity), convert the methane to methyl alcohol, or sent to natural gas lines after cleaning process [188].

5.2. Biopulping and wood utilization

Biopulping, also known as biological pulping, refers a type of industrial biotechnology using fungus to convert wood chips to paper pulp. This technology has the potential to improve the quality of paper pulp, reduce energy consumption and environmental impacts when compare with the traditional chemical pulping technologies [189].

The aim of pulping is to extract cellulose from plant material. The traditional approaches are mechanical and chemical pulping. The former method is generally accomplished by refining grinding or thermo-mechanical pulping. The latter way is to dissolve lignin from the cellulose and hemicellulose fibers via chemical treatment, such as kraft pulping in which wood chips are cooled in a solution containing sodium hydroxide and sodium sulfide [190]. These traditional pulping technologies have several drawbacks: (1) high energy demand; (2) low cellulose yield, especially from chemical pulping due to partial degradation of cellulose; (3) potential hazards chemicals emitted to the environment [189].

Lignin is a complex polymer which serves as a structural component of higher plants and is highly resistant towards chemical degradation [191]. White-rot and brown-rot fungi are two classifications of wood-rotting basidiomycetes. White-rot basidimycetes have been reported enable to, selectively or simultaneously with cellulose, degrade lignin in different types of wood [191, 192]. Brown-rot basidiomycetes, which grow mainly on softwood, can degrade wood polysaccharides but cause only a partial modification of lignin. Besides white- and brown- rot basidimycetes, some scomycetes so-called soft-rot fungi which can degrade wood under extreme environmental conditions such as high or low water potential that prohibit the activity of other fungi [191].

The fungal treatment process fits in a paper mill operation well. After wood is debarked, chipped and screened, wood chips are briefly steamed to reduce natural chip microorganisms, cooled with air, and inoculated with the biopulping fungus for 1 to 4 weeks prior to further processing. The biopulping has been indicated as a technology technologically feasible and economically beneficial [193].

This biological treatment of wood using fungi has also been studied and used as a pre-treatment approach prior to enzymatic hydrolysis for biofuel production [194-196]. However, more research are required to understand the mechanism of wood degradation, structural changes of wood cell wall caused by these wood decay fungus and to improve the treatment technologies [197, 198].

6. Conclusions

The concept of 'biorefinery' has emerged since the potential of lignocellulosic based products substituting fossil fuel derived products has been discovered. Biorefienries may play a major role in tackling climate change by reducing the demand on fossil fuel energy and providing sustainable energy, chemicals and materials, potentially aiding energy security, and creating opportunities and market. This paper reviewed a wide range of such lignocellulosic derived products and current available biorefinery technologies. Some of these technologies have been or being close to the industrialization and others are still at the early stage of development. However, more research efforts are required to improve the technologies and integrate the biorefinery system in order to achieve the maximum outputs and to make biorefinery work at scale.

Author details

Hongbin Cheng*
Department of Process Engineering, Stellenbosch University, South Africa
New China Times Technology Ltd, China

Lei Wang*
Department of Life Science, Imperial College London, UK
New China Times Technology Ltd, China

7. References

[1] Clark, J.H., Green chemistry for the second generation biorefinery—sustainable chemical manufacturing based on biomass. Journal of Chemical Technology & Biotechnology, 2007. 82(7): p. 603-609.

* The two authors contribute equally to this study

[2] Stöcker, M., *Biofuels and Biomass-To-Liquid Fuels in the Biorefinery: Catalytic Conversion of Lignocellulosic Biomass using Porous Materials.* Angewandte Chemie International Edition, 2008. 47(48): p. 9200-9211.

[3] Gullón, P., et al., *Selected Process Alternatives for Biomass Refining: A Review.* The Open Agricultrure Journal 2010. 4: p. 135-144.

[4] Kelley, S.S., *Lignocellulosic Biorefineries: Reality, Hype, or Something in Between?*, in *Materials, Chemicals, and Energy from Forest Biomass.* 2007, American Chemical Society. p. 31-47.

[5] Danner, H. and R. Braun, *ChemInform Abstract: Biotechnology for the Production of Commodity Chemicals from Biomass.* ChemInform, 2000. 31(6): p. no-no.

[6] Chheda, J.N., G.W. Huber, and J.A. Dumesic, *Liquid-Phase Catalytic Processing of Biomass-Derived Oxygenated Hydrocarbons to Fuels and Chemicals.* Angewandte Chemie International Edition, 2007. 46(38): p. 7164-7183.

[7] Rowlands, W.N., A. Masters, and T. Maschmeyer, *The Biorefinery—Challenges, Opportunities, and an Australian Perspective.* Bulletin of Science, Technology & Society, 2008. 28(2): p. 149-158.

[8] Sanders, J.P.M., B. Annevelink, and D.A.v.d. Hoeven, *The development of biocommodities and the role of North West European ports in biomass chains.* Biofuels Bioproducts and Biorefining, 2009. 3(3): p. 395-409.

[9] Kamm, B., *Production of Platform Chemicals and Synthesis Gas from Biomass.* Angewandte Chemie International Edition, 2007. 46(27): p. 5056-5058.

[10] Kamm, B., P. Schönicke, and M. Kamm, *Biorefining of Green Biomass – Technical and Energetic Considerations.* CLEAN – Soil, Air, Water, 2009. 37(1): p. 27-30.

[11] Li, L., S. Lu, and V. Chiang, *A Genomic and Molecular View of Wood Formation.* Critical Reviews in Plant Sciences, 2006. 25(3): p. 215-233.

[12] Kamm, B., et al., *Biorefinery Systems – An Overview*, in *Biorefineries-Industrial Processes and Products.* 2008, Wiley-VCH Verlag GmbH. p. 1-40.

[13] Fernando, S., et al., *Biorefineries: Current Status, Challenges, and Future Direction.* Energy & Fuels, 2006. 20(4): p. 1727-1737.

[14] Kamm, B. and M. Kamm, *Biorefinery - Systems.* Chemical and Biochemical Engineering Quarterly, 2004. 18(1): p. 1-6.

[15] Mosier, N., et al., *Features of promising technologies for pretreatment of lignocellulosic biomass.* Bioresource Technology, 2005. 96(6): p. 673-686.

[16] Zheng, Y., Z. Pan, and R. Zhang, *Overview of biomass pretreatment for cellulosic ethanol production.* International Journal of Agricultural and Biological Engineering, 2009. 2(3): p. 51-68.

[17] McKendry, P., *Energy production from biomass (part 1): overview of biomass.* Bioresource Technology, 2002. 83(1): p. 37-46.

[18] Kristian, M. and M. Hurme, *Lignocellulosic biorefinery economic evaluation.* Cellulose Chemistry and Technology, 2011. 45(7-8): p. 443-454.

[19] Lynn, W., *Biomass Energy Data Book: Edition 1.* 2006.

[20] Wyman, C. and B. Goodman, *Biotechnology for production of fuels, chemicals, and materials from biomass.* Applied Biochemistry and Biotechnology, 1993. 39-40(1): p. 41-59.

[21] García, A., et al., *Biorefining of lignocellulosic residues using ethanol organosolv process.* Chemical Engineering Transactions, 2009. 18: p. 911-916.

[22] Puppan, D., *Environmental evaluation of biofuels.* Periodica Polytechnica Ser. Soc. Man. Sci., 2002. 10(1): p. 95-116.

[23] Lin, Y. and S. Tanaka, *Ethanol fermentation from biomass resources: current state and prospects.* Applied Microbiology and Biotechnology, 2006. 69(6): p. 627-642.

[24] Slade, R.B., *Prospects for cellulosic ethanol supply-chains in Europe. a techno-economic and environmental assessment,* in *Centre for Process Systems Engineering and Centre for Environmental Policy.* 2009, University of London. p. 170.

[25] Josefsson, T., H. Lennholm, and G. Gellerstedt, *Steam Explosion of Aspen Wood. Characterisation of Reaction Products.* Holzforschung, 2002. 56(3): p. 289-297.

[26] Ballesteros, I., et al., *Effect of chip size on steam explosion pretreatment of softwood.* Applied Biochemistry and Biotechnology, 2000. 84-86(1): p. 97-110.

[27] Ruiz, E., et al., *Evaluation of steam explosion pre-treatment for enzymatic hydrolysis of sunflower stalks.* Enzyme and Microbial Technology, 2008. 42(2): p. 160-166.

[28] Holtzapple, M., et al., *Pretreatment of lignocellulosic municipal solid waste by ammonia fiber explosion (AFEX).* Applied Biochemistry and Biotechnology, 1992. 34 35(1): p. 5 21.

[29] Alizadeh, H., et al., *Pretreatment of switchgrass by ammonia fiber explosion (AFEX).* Applied Biochemistry and Biotechnology, 2005. 124(1): p. 1133-1141.

[30] Tengborg, C., et al., *Comparison of SO2 and H2SO4 impregnation of softwood prior to steam pretreatment on ethanol production.* Applied Biochemistry and Biotechnology, 1998. 70-72(1): p. 3-15.

[31] Öhgren, K., M. Galbe, and G. Zacchi, *Optimization of Steam Pretreatment of SO2;-Impregnated Corn Stover for Fuel Ethanol Production,* in *Twenty-Sixth Symposium on Biotechnology for Fuels and Chemicals,* B.H. Davison, et al., Editors. 2005, Humana Press. p. 1055-1067.

[32] Kim, K.H. and J. Hong, *Supercritical CO2 pretreatment of lignocellulose enhances enzymatic cellulose hydrolysis.* Bioresource Technology, 2001. 77(2): p. 139-144.

[33] Zheng, Y., H.M. Lin, and G.T. Tsao, *Pretreatment for Cellulose Hydrolysis by Carbon Dioxide Explosion.* Biotechnology Progress, 1998. 14(6): p. 890-896.

[34] Mosier, N., et al., *Optimization of pH controlled liquid hot water pretreatment of corn stover.* Bioresource Technology, 2005. 96(18): p. 1986-1993.

[35] Laser, M., et al., *A comparison of liquid hot water and steam pretreatments of sugar cane bagasse for bioconversion to ethanol.* Bioresource Technology, 2002. 81(1): p. 33-44.

[36] Allen, S.G., et al., *A Comparison between Hot Liquid Water and Steam Fractionation of Corn Fiber.* Industrial & Engineering Chemistry Research, 2001. 40(13): p. 2934-2941.

[37] Aden, A., Ruth, M., Ibsen, K., Jechura, J., Neeves, K., Sheehan, J., Wallace, B., Montague, L., Slayton, A., Lukas, J, *Lignocellulosic Biomass to Ethanol Process Design and Economics Utilizing Co-Current Dilute Acid Prehydrolysis and Enzymatic Hydrolysis for Corn Stover.* 2002, National Renewable Energy Laboratory (NREL). p. 95. NREL/TP-510-32438.

[38] Torget, R., M.E. Himmel, and K. Grohmann, *Dilute sulfuric acid pretreatment of hardwood bark.* Bioresource Technology, 1991. 35(3): p. 239-246.

[39] Schell, D.J., et al., *Dilute-sulfuric acid pretreatment of corn stover in pilot-scale reactor - Investigation of yields, kinetics, and enzymatic digestibilities of solids.* Applied Biochemistry and Biotechnology, 2003. 105: p. 69-85.

[40] Kim, T.H. and Y.Y. Lee, *Fractionation of corn stover by hot-water and aqueous ammonia treatment.* Bioresource Technology, 2006. 97(2): p. 224-232.

[41] Vaccarino, C., et al., *Effect of SO2, NaOH and Na2CO3 pretreatments on the degradability and cellulase digestibility of grape marc.* Biological Wastes, 1987. 20(2): p. 79-88.

[42] Chander Kuhad, R., et al., *Fed batch enzymatic saccharification of newspaper cellulosics improves the sugar content in the hydrolysates and eventually the ethanol fermentation by Saccharomyces cerevisiae.* Biomass and Bioenergy, 2010. 34(8): p. 1189-1194.

[43] Sierra, R., L.A. Garcia, and M.T. Holtzapple, *Selectivity and delignification kinetics for oxidative short-term lime pretreatment of poplar wood. Part I: Constant-pressure.* Biotechnology Progress, 2011. 27(4): p. 976-985.

[44] Chang, V., et al., *Oxidative lime pretreatment of high-lignin biomass.* Applied Biochemistry and Biotechnology, 2001. 94(1): p. 1-28.

[45] Pasquini, D., et al., *Extraction of lignin from sugar cane bagasse and Pinus taeda wood chips using ethanol–water mixtures and carbon dioxide at high pressures.* The Journal of Supercritical Fluids, 2005. 36(1): p. 31-39.

[46] Hallac, B.B., et al., *Effect of Ethanol Organosolv Pretreatment on Enzymatic Hydrolysis of Buddleja davidii Stem Biomass.* Industrial & Engineering Chemistry Research, 2010. 49(4): p. 1467-1472.

[47] Brandt, A., et al., *Ionic liquid pretreatment of lignocellulosic biomass with ionic liquid-water mixtures.* Green Chemistry, 2011. 13(9): p. 2489-2499.

[48] Hamelinck, C.N., G.v. Hooijdonk, and A.P.C. Faaij, *Ethanol from lignocellulosic biomass: techno-economic performance in short-, middle- and long-term.* Biomass and Bioenergy, 2005. 28(4): p. 384-410.

[49] Sedlak, M. and N. Ho, *Production of ethanol from cellulosic biomass hydrolysates using genetically engineered saccharomyces yeast capable of cofermenting glucose and xylose.* Applied biochemistry and biotechnology, 2004. 114(1): p. 403-416.

[50] Humbird, D. and A. Aden, *Biochemical Production of Ethanol from Corn Stover: 2008 State of Technology Model.* 2009, National Renewable Energy Laboratory (NREL). NREL/TP-510-46214.

[51] Cripps, R.E., et al., *Metabolic engineering of Geobacillus thermoglucosidasius for high yield ethanol production.* Metabolic Engineering, 2009. 11(6): p. 398-408.

[52] Bezerra, R. and A. Dias, *Enzymatic kinetic of cellulose hydrolysis: Inhibition by ethanol and cellobiose.* Applied Biochemistry and Biotechnology, 2005. 126(1): p. 49-59.

[53] Lynd, L.R., et al., *Consolidated bioprocessing of cellulosic biomass: an update.* Current Opinion in Biotechnology, 2005. 16: p. 577-583.

[54] van Zyl, W., et al., *Consolidated Bioprocessing for Bioethanol Production Using Biofuels*, L. Olsson, Editor. 2007, Springer Berlin / Heidelberg. p. 205-235.

[55] Jin, M., et al., *Consolidated bioprocessing (CBP) performance of Clostridium phytofermentans on AFEX-treated corn stover for ethanol production.* Biotechnology and Bioengineering, 2011. 108(6): p. 1290-1297.

[56] Olson, D.G., et al., *Recent progress in consolidated bioprocessing.* Current Opinion in Biotechnology, (0).

[57] Phillips, S., et al., *Thermochemical Ethanol via Indirect Gasification and Mixed Alcohol Synthesis of Lignocellulosic Biomass.* 2007, National Renewable Energy Laboratory (NREL). NREL/TP-510-41168.

[58] Dutta, A., et al., *An economic comparison of different fermentation configurations to convert corn stover to ethanol using Z. mobilis and Saccharomyces.* Biotechnology Progress, 2010. 26(1): p. 64-72.

[59] Stephenson, A.L., et al., *The environmental and economic sustainability of potential bioethanol from willow in the UK.* Bioresource Technology, 2010. 101(24): p. 9612-9623.

[60] Sassner, P., M. Galbe, and G. Zacchi, *Techno-economic evaluation of bioethanol production from three different lignocellulosic materials.* Biomass and Bioenergy, 2008. 32(5): p. 422-430.

[61] Wang, L., et al., *Technology performance and economic feasibility of bioethanol production from various waste papers.* Energy & Environmental Science 2012. 5(2): p. 5717-5730.

[62] Mu, D., et al., *Comparative Life Cycle Assessment of Lignocellulosic Ethanol Production: Biochemical Versus Thermochemical Conversion.* Environmental Management, 2010. 46(4): p. 565-578.

[63] Qureshi, N. and T.C. Ezeji, *Butanol, 'a superior biofuel' production from agricultural residues (renewable biomass): recent progress in technology.* Biofuels, Bioproducts and Biorefining, 2008. 2(4): p. 319-330.

[64] Ezeji, T., N. Qureshi, and H.P. Blaschek, *Butanol production from agricultural residues: Impact of degradation products on Clostridium beijerinckii growth and butanol fermentation.* Biotechnology and Bioengineering, 2007. 97(6): p. 1460-1469.

[65] Cascone, R., *Biobutanol – a replacement for bioethanol?* Chemical Engineering Progressing, 2008. 104: p. S4-S9.

[66] Pfromm, P.H., et al., *Bio-butanol vs. bio-ethanol: A technical and economic assessment for corn and switchgrass fermented by yeast or Clostridium acetobutylicum.* Biomass and Bioenergy, 2010. 34(4): p. 515-524.

[67] Shereena, K.M. and T. Thangaraj, *Biodiesel: an alternative fuel produced from vegetable oils by transesterification.* Electronic Journal of Biology, 2009. 5(3): p. 67-74.

[68] Zabeti, M., W.M.A. Wan Daud, and M.K. Aroua, *Activity of solid catalysts for biodiesel production: A review.* Fuel Processing Technology, 2009. 90(6): p. 770-777.

[69] Ma, F. and M.A. Hanna, *Biodiesel production: a review.* Bioresource Technology, 1999. 70(1): p. 1-15.

[70] Marchetti, J.M., V.U. Miguel, and A.F. Errazu, *Possible methods for biodiesel production.* Renewable and Sustainable Energy Reviews, 2007. 11(6): p. 1300-1311.

[71] Schnepf, R., *Agriculture-Based Biofuels: Overview and Emerging Issues.* 2010, Congressional Research Service. R41282.

[72] Tamalampudi, S., et al., *Enzymatic production of biodiesel from Jatropha oil: A comparative study of immobilized-whole cell and commercial lipases as a biocatalyst.* Biochemical Engineering Journal, 2008. 39(1): p. 185-189.

[73] Lu, H., et al., *Production of biodiesel from Jatropha curcas L. oil.* Computers & Chemical Engineering, 2009. 33(5): p. 1091-1096.

[74] Shah, S. and M.N. Gupta, *Lipase catalyzed preparation of biodiesel from Jatropha oil in a solvent free system*. Process Biochemistry, 2007. 42(3): p. 409-414.

[75] Kumari, A., et al., *Enzymatic transesterification of Jatropha oil*. Biotechnology for Biofuels, 2009. 2(1): p. 1.

[76] Chisti, Y., *Biodiesel from microalgae*. Biotechnology Advances, 2007. 25: p. 294-306.

[77] Scott, S.A., et al., *Biodiesel from algae: challenges and prospects*. Current Opinion in Biotechnology, 2010. 21(3): p. 277-286.

[78] Grobbelaar, J., *Turbulence in mass algal cultures and the role of light/dark fluctuations*. Journal of Applied Phycology, 1994. 6(3): p. 331-335.

[79] Chisti, Y., *Biodiesel from microalgae beats bioethanol*. Trends in Biotechnology, 2008. 26(3): p. 126-131.

[80] Monson, K.D., et al., *Anaerobic digestion of biodegrable municiple solid wastes: A review*. 2007, University of Glamorgan

[81] Trzcinski, A.P., *Anaerobic membrane bioreactor technology for solid waste stabilization*, in *Chemical Engineering and Chemical Technology*. 2009, Imperial College London: London.

[82] Rapport, J., et al., *Current anaerobic digestion technologies used for treatment of municipal organic solid waste*. 2008, Department of Biological and Agricultural Engineering, University of California Davis, CA

[83] Raposo, F., et al., *Influence of inoculum to substrate ratio on the biochemical methane potential of maize in batch tests*. Process Biochemistry, 2006. 41(6): p. 1444-1450.

[84] Hartmann, H., H.B. Moller, and B.K. Ahring, *Efficiency of the anaerobic treatment of the organic fraction of municipal solid waste: collection and pretreatment*. Waste Management & Research, 2004. 22: p. 35-41.

[85] WRAP, *Anaerobic digestate*. 2008, Waste & Resources Action Programme (WRAP)

[86] Edelmann, W., K. Schleiss, and A. Joss, *Ecological, energetic and economic comparison of anaerobic digestion with different competing technologies to treat biogenic wastes*. Water Science and Technology, 2000. 4(3): p. 263-273.

[87] Edelmann, W., U. Baier, and H. Engeli, *Environmental aspects of the anaerobic digestion of the organic fraction of municipal solid wastes and of solid agricultural wastes*. Water Science and Technology, 2004. 52(1-2): p. 203-208.

[88] Dodds, D.R. and R.A. Gross, *Chemicals from Biomass*. Science, 2007. 318(5854): p. 1250-1251.

[89] Carlson, T.L. and E.M. Peters. 2006. Patent application US2006/094093 A1.

[90] Lorenz, P. and H. Zinke, *White biotechnology: differences in US and EU approaches?* Trends in Biotechnology, 2005. 23(12): p. 570-574.

[91] Jung, Y.K., et al., *Metabolic engineering of Escherichia coli for the production of polylactic acid and its copolymers*. Biotechnology and Bioengineering, 2010. 105(1): p. 161-171.

[92] Wyman, C.E., et al., *Coordinated development of leading biomass pretreatment technologies*. Bioresource Technology, 2005. 96(18): p. 1959-1966.

[93] Yáñez, R., et al., *Production of D(−)-lactic acid from cellulose by simultaneous saccharification and fermentation using <i>Lactobacillus coryniformis</i> subsp. <i>torquens</i>*. Biotechnology Letters, 2003. 25(14): p. 1161-1164.

[94] Adsul, M.G., et al., *Development of biocatalysts for production of commodity chemicals from lignocellulosic biomass*. Bioresource Technology, 2011. 102(6): p. 4304-4312.

[95] Danner, H., et al., *Bacillus stearothermophilus for thermophilic production of l-lactic acid*. Applied Biochemistry and Biotechnology, 1998. 70-72(1): p. 895-903.

[96] Maddox, I.S., N. Qureshi, and K. Roberts-Thomson, *Production of acetone-butanol-ethanol from concentrated substrate using clostridium acetobutylicum in an integrated fermentation-product removal process*. Process Biochemistry, 1995. 30(3): p. 209-215.

[97] Jones, D.T. and D.R. Woods, *Acetone-butanol fermentation revisited*. Microbiology and Molecular Biology Reviews, 1986. 50(4): p. 484-524.

[98] Falk, A., *Xylan biosynthesis: News from the grass*. Plant Physiology, 2010. 153: p. 396-402.

[99] Granstrom, T., K. Izumori, and M. Leisola, *A rare sugar xylitol. Part II: biotechnological production and future applications of xylitol*. Applied Microbiology and Biotechnology, 2007. 74(2): p. 273-276.

[100] Saha, B., *Hemicellulose bioconversion*. Journal of Industrial Microbiology & Biotechnology, 2003. 30(5): p. 279-291.

[101] Choct, M. and G. Annison, *Anti-nutritive activity of wheat pentosans in broiler diets*. British Poultry Science, 1990. 31(4): p. 811-821.

[102] Duff, S.J.B. and W.D. Murray, *Bioconversion of forest products industry waste cellulosics to fuel ethanol: a review*. Bioresource Technology, 1996. 55: p. 1-33.

[103] Hayes, D.J., et al., *The Biofine Process – Production of Levulinic Acid, Furfural, and Formic Acid from Lignocellulosic Feedstocks*, in *Biorefineries-Industrial Processes and Products*. 2008, Wiley-VCH Verlag GmbH. p. 139-164.

[104] Fatih Demirbas, M., *Biorefineries for biofuel upgrading: A critical review*. Applied Energy, 2009. 86, Supplement 1(0): p. S151-S161.

[105] Saddler, J.N., *Bioconversion of Forest and Agricultural Plant Residues*. Biotechnology in Agriculture 1993. CABI, Wallingford, UK

[106] Sun, Y. and J.Y. Cheng, *Hydrolysis of lignocellulosic materials for ethanol production: a review*. Bioresource Technology, 2002. 83(1): p. 1-11.

[107] Zhu, S., et al., *FED-BATCH SIMULTANEOUS SACCHARIFICATION AND FERMENTATION OF MICROWAVE/ACID/ALKALI/H2O2 PRETREATED RICE STRAW FOR PRODUCTION OF ETHANOL*. Chemical Engineering Communications, 2006. 193(5): p. 639-648.

[108] Koukios, E.G. and G.N. Valkanas, *Process for chemical separation of the three main components of lignocellulosic biomass*. Industrial & Engineering Chemistry Product Research and Development, 1982. 21(2): p. 309-314.

[109] Koukios, E.G., *Biomass refining: a non-waste approach.*, in *Economics and Ecosystem Management*, D.O. Hall, N. Myers, and N.S. Margaris, Editors. 1985, Dr W. Junk Publishers: Dordrecht, The Netherlands. p. 233-244.

[110] Kamm, B., P.R. Gruber, and M. Kamm, *Biorefineries – Industrial Processes and Products*, in *Ullmann's Encyclopedia of Industrial Chemistry*. 2000, Wiley-VCH Verlag GmbH & Co. KGaA.

[111] Kleinert, T.N., *Organosolv pulping with aqueous alcohol*. Tappi Journal, 1974. 57: p. 99-102.

[112] Lora, J.H. and S. Aziz, *Organosolv pulping — a versatile approach to wood refining.* Tappi Journal, 1985. 68(6): p. 94-97.

[113] Sarkanen, K.V., *Chemistry of solvent pulping.* Tappi Journal, 1990: p. 215-219.

[114] Pan, X., et al., *Bioconversion of hybrid poplar to ethanol and co-products using an organosolv fractionation process: Optimization of process yields.* Biotechnology and Bioengineering, 2006. 94(5): p. 851-861.

[115] Pan, X., et al., *Effect of organosolv ethanol pretreatment variables on physical characteristics of hybrid poplar substrates.* Applied Biochemistry and Biotechnology, 2007. 137-140(1): p. 367-377.

[116] Pan, X., et al., *The bioconversion of mountain pine beetle-killed lodgepole pine to fuel ethanol using the organosolv process.* Biotechnology and Bioengineering, 2008. 101(1): p. 39-48.

[117] Taherzadeh, M.J. and K. Karimi, *Pretreatment of lignocellulosic wastes to improve ethanol and biogas production: A review.* International Journal of Molecular Sciences, 2008. 9(9): p. 1621-1651.

[118] Chum, H.L., et al., *Organosolv pretreatment for enzymatic hydrolysis of poplars: I. Enzyme hydrolysis of cellulosic residues.* Biotechnology Bioengineering, 1988. 31(7).

[119] Zhao, X., K. Cheng, and D. Liu, *Organosolv pretreatment of lignocellulosic biomass for enzymatic hydrolysis.* Applied Microbiology and Biotechnology, 2009. 82(5): p. 815-827.

[120] Johansson, A., O. Aaltonen, and P. Ylinen, *Organosolv pulping — methods and pulp properties.* Biomass, 1987. 13(1): p. 45-65.

[121] Aziz, S. and K.V. Sarkanen, *Organosolv pulping—a review.* Tappi Journal, 1989. 72: p. 169-175.

[122] Thring, R.W., E. Chornet, and R.P. Overend, *Recovery of a solvolytic lignin: Effects of spent liquor/acid volume ratio, acid concentration and temperature.* Biomass, 1990. 23(4): p. 289-305.

[123] Sidiras, D. and E. Koukios, *Simulation of acid-catalysed organosolv fractionation of wheat straw.* Bioresource Technology, 2004. 94(1): p. 91-98.

[124] Zhu, J.Y. and X.J. Pan, *Woody biomass pretreatment for cellulosic ethanol production: Technology and energy consumption evaluation.* Bioresource Technology, 2010. 101(13): p. 4992-5002.

[125] Myllymaki, V. and R. Aksela. 2005. Dissolution and delignification of lignocellulosic materials with ionic liquid solvent under microwave irradiation. WO patent 2005/017001

[126] Hayes, D.J., *An examination of biorefining processes, catalysts and challenges.* Catalysis Today, 2009. 145(1–2): p. 138-151.

[127] Dadi, A.P., S. Varanasi, and C.A. Schall, *Enhancement of cellulose saccharification kinetics using an ionic liquid pretreatment step.* Biotechnology and Bioengineering, 2006. 95(5): p. 904-910.

[128] Zhu, S., *Use of ionic liquids for the efficient utilization of lignocellulosic materials.* Journal of Chemical Technology & Biotechnology, 2008. 83(6): p. 777-779.

[129] Reichert, W.M., et al., *Derivatization of chitin in room temperature ionic liquids,* in *222 ACS National Meeting.* 2001. Chicago, IL, United States.

[130] Heinze, T., K. Schwikal, and S. Barthel, *Ionic Liquids as Reaction Medium in Cellulose Functionalization.* Macromolecular Bioscience, 2005. 5(6): p. 520-525.

[131] Dien, B.S., et al., *Enzymatic saccharification of hot-water pretreated corn fiber for production of monosaccharides.* Enzyme and Microbial Technology, 2006. 39(5): p. 1137-1144.

[132] Negro, M., et al., *Hydrothermal pretreatment conditions to enhance ethanol production from poplar biomass.* Applied Biochemistry and Biotechnology, 2003. 105(1): p. 87-100.

[133] Rogalinski, T., T. Ingram, and G. Brunner, *Hydrolysis of lignocellulosic biomass in water under elevated temperatures and pressures.* The Journal of Supercritical Fluids, 2008. 47(1): p. 54-63.

[134] van Walsum, G., et al., *Conversion of lignocellulosics pretreated with liquid hot water to ethanol.* Applied Biochemistry and Biotechnology, 1996. 57-58(1): p. 157-170.

[135] Franck, E.U., *Fluids at high pressures and temperatures.* Chemical thermodynamics, 1987. 19: p. 225-242.

[136] Shaw, R.W., et al., *Supercritical water. A medium for chemistry.* Chemical Engineering News, 1991. 23: p. 26-39.

[137] Savage, P.E., et al., *Reactions at supercritical conditions: [A]pplications and fundamentals.* AIChE Journal, 1995. 41(7): p. 1723-1778.

[138] Katritzky, A.R., et al., *Reactions in High-Temperature Aqueous Media.* Chemical Reviews, 2001. 101(4): p. 837-892.

[139] Kohlmann, K.L., et al., *Enhanced Enzyme Activities on Hydrated Lignocellulosic Substrates*, in *Enzymatic Degradation of Insoluble Carbohydrates.* 1996, American Chemical Society. p. 237-255.

[140] Weil, J., et al., *Continuous pH monitoring during pretreatment of yellow poplar wood sawdust by pressure cooking in water.* Applied Biochemistry and Biotechnology, 1998. 70-72(1): p. 99-111.

[141] Alvira, P., et al., *Pretreatment technologies for an efficient bioethanol production process based on enzymatic hydrolysis: A review.* Bioresource Technology, 2010. 101(13): p. 4851-4861.

[142] Goring, D.A.I., *Thermal softening of lignin, hemicelluloses and cellulose.* 1963: Pulp and Paper Research Institute of Canada.

[143] Overend, R.P., E. Chornet, and J.A. Gascoigne, *Fractionation of Lignocellulosics by Steam-Aqueous Pretreatments [and Discussion].* Philosophical Transactions of the Royal Society of London. Series A, Mathematical and Physical Sciences, 1987. 321(1561): p. 523-536.

[144] Mason, W.H. 1929. Apparatus and process of explosion defibration of lignocellulosic material. U.S. Patent No. 1655618

[145] Montane, D., et al., *Application of steam explosion to the fractionation and rapid vapor-phase alkaline pulping of wheat straw.* Biomass and Bioenergy, 1998. 14(3): p. 261-276.

[146] Pan, X., et al., *Strategies to enhance the enzymatic hydrolysis of pretreated softwood with high residual lignin content.* Applied Biochemistry and Biotechnology, 2005. 121-124: p. 1069-1079.

[147] Eklund, R., M. Galbe, and G. Zacchi, *The influence of SO2 and H2SO4 impregnation of willow prior to steam pretreatment.* Bioresource Technology, 1995. 52(3): p. 225-229.

[148] Fein, J.E., D. Potts, and D. Good, *Development of an optimal wood-to-fuel ethanol process utilizing best available technology.* Energy Biomass and Waste, 1991. 15(745-765).

[149] Ropars, M., et al., *Large-scale enzymatic hydrolysis of agricultural lignocellulosic biomass. Part 1: Pretreatment procedures.* Bioresource Technology, 1992. 42(3): p. 197-204.

[150] Morjanoff, P.J. and P.P. Gray, *Optimization of steam explosion as method for increasing susceptibility of sugarcane bagasse to enzymatic saccharification.* Biotechnology Bioengineering, 1987. 29(6): p. 733-741.

[151] Focher, B., A. Marzetti, and V. Crescenzi, *Steam Explosion Techniques, Fundamentals and Industrial Applications.* 1991, Gordon and Breach Publishers: Philadelphia. p. 412.

[152] Avellar, B.K. and W.G. Glasser, *Steam-assisted biomass fractionation. I. Process considerations and economic evaluation.* Biomass and Bioenergy, 1998. 14(3): p. 205-218.

[153] Zheng, Y. and G.T. Tsao, *Avicel hydrolysis by cellulase enzyme in supercritical CO2.* Biotechnology Letters, 1996. 18(4): p. 451-454.

[154] Zheng, Y., et al., *Supercritical carbon dioxide explosion as a pretreatment for cellulose hydrolysis.* Biotechnology Letters, 1995. 17(8): p. 845-850.

[155] Schacht, C., C. Zetzl, and G. Brunner, *From plant materials to ethanol by means of supercritical fluid technology.* The Journal of Supercritical Fluids, 2008. 46(3): p. 299-321.

[156] Laureano-Perez, L., et al., *Understanding Factors that Limit Enzymatic Hydrolysis of Biomass*

[157] *nty-Sixth Symposium on Biotechnology for Fuels and Chemicals,* B.H. Davison, et al., Editors. 2005, Humana Press. p. 1081-1099.

[158] Teymouri, F., et al., *Optimization of the ammonia fiber explosion (AFEX) treatment parameters for enzymatic hydrolysis of corn stover.* Bioresource Technology, 2005. 96(18): p. 2014-2018.

[159] Eggeman, T. and R.T. Elander, *Process and economic analysis of pretreatment technologies.* Bioresource Technology, 2005. 96(18): p. 2019-2025.

[160] Ghosh, S., et al., *Pilot-scale gasification of municipal solid wastes by high-rate and two-phase anaerobic digestion (TPAD).* Water Science and Technology 2000. 41(3): p. 101-110.

[161] Beccari, M., et al., *Enhancement of anaerobic treatability of olive oil mill effluents by addition of Ca(OH)2 and bentonite without intermediate solid/liquid separation.* Water Science and Technology, 2001. 43(11): p. 275-282.

[162] Tanaka, S., et al., *Effects of thermochemical pretreatment on the anaerobic digestion of waste activated sludge.* Water Science and Technology, 1997. 35(8): p. 209-215.

[163] Brownell, H.H., E.K.C. Yu, and J.N. Saddler, *Steam-explosion pretreatment of wood: Effect of chip size, acid, moisture content and pressure drop.* Biotechnology and Bioengineering, 1986. 28(6): p. 792-801.

[164] Gregg, D. and J. Saddler, *A techno-economic assessment of the pretreatment and fractionation steps of a biomass-to-ethanol process.* Applied Biochemistry and Biotechnology, 1996. 57-58(1): p. 711-727.

[165] Chang, V.S., et al., *Simultaneous saccharification and fermentation of lime-treated biomass.* Biotechnology Letters, 2001. 23(16): p. 1327-1333.

[166] Sun, X.F., et al., *Characteristics of degraded cellulose obtained from steam-exploded wheat straw.* Carbohydrate Research, 2005. 340(1): p. 97-106.

[167] Jones, J.L. and K.T. Semrau, *Wood hydrolysis for ethanol production — previous experience and the economics of selected processes*. Biomass, 1984. 5(2): p. 109-135.

[168] Kootstra, A.M.J., et al., *Comparison of dilute mineral and organic acid pretreatment for enzymatic hydrolysis of wheat straw*. Biochemical Engineering Journal, 2009. 46(2): p. 126-131.

[169] Bjerre, A.B., et al., *Pretreatment of wheat straw using combined wet oxidation and alkaline hydrolysis resulting in convertible cellulose and hemicellulose*. Biotechnology and Bioengineering, 1996. 49(5): p. 568-577.

[170] Kumar, P., et al., *Methods for Pretreatment of Lignocellulosic Biomass for Efficient Hydrolysis and Biofuel Production*. Industrial & Engineering Chemistry Research, 2009, 48(8). p. 3713-3729.

[171] Carvalheiro, F., L.C. Duarte, and F.M. Girio, *Hemicellulose biorefineries: a review on biomass pretreatments*. Scientific & Industrial Research, 2008. 67: p. 849-864.

[172] Molero Gómez, A., C. Pereyra López, and E. Martinez de la Ossa, *Recovery of grape seed oil by liquid and supercritical carbon dioxide extraction: a comparison with conventional solvent extraction*. The Chemical Engineering Journal and the Biochemical Engineering Journal, 1996. 61(3). p. 227-231.

[173] Toews, K.L., et al., *pH-Defining Equilibrium between Water and Supercritical CO2. Influence on SFE of Organics and Metal Chelates*. Analytical Chemistry, 1995. 67(22): p. 4040-4043.

[174] Intanakul, P., M. Krairiksh, and P. Kitchaiya, *Enhancement of Enzymatic Hydrolysis of Lignocellulosic Wastes by Microwave Pretreatment Under Atmospheric Pressure*. Journal of Wood Chemistry and Technology, 2003. 23(2): p. 217-225.

[175] Saha, B.C., A. Biswas, and M.A. Cotta, *Microwave Pretreatment, Enzymatic Saccharification and Fermentation of Wheat Straw to Ethanol*. Journal of Biobased Materials and Bioenergy, 2008. 2(3) p. 210-217.

[176] Ma, H., et al., *Enhanced enzymatic saccharification of rice straw by microwave pretreatment*. Bioresource Technology, 2009. 100(3): p. 1279-1284.

[177] Imai, M., K. Ikari, and I. Suzuki, *High-performance hydrolysis of cellulose using mixed cellulase species and ultrasonication pretreatment*. Biochemical Engineering Journal, 2004. 17(2): p. 79-83.

[178] Nitayavardhana, S., et al., *Ultrasound pretreatment of cassava chip slurry to enhance sugar release for subsequent ethanol production*. Biotechnology and Bioengineering, 2008. 101(3): p. 487-496.

[179] Keshwani, D.R., *Microwave Pretreatment of Switchgrass for Bioethanol Production*. 2009, North Carolina State University.

[180] Sun, R.C. and J. Tomkinson, *Characterization of hemicelluloses obtained by classical and ultrasonically assisted extractions from wheat straw*. Carbohydrate Polymers, 2002. 50(3): p. 263-271.

[181] Yachmenev, V., et al., *Acceleration of the enzymatic hydrolysis of corn stover and sugar cane bagasse celluloses by low intensity uniform ultrasound*. Biobased Material Bioenergy, 2009. 3: p. 25-31.

[182] Vinatoru, M.e.a., *Ultrasonically assisted extraction of bioactive principles from plants and their constituents*, in *Advances in Sonochemistry*, T.J. Mason, Editor. 1999, JAI Press. p. 209-248.

[183] Palonen, H., et al., *Evaluation of wet oxidation pretreatment for enzymatic hydrolysis of softwood*. Applied Biochemistry and Biotechnology, 2004. 117(1): p. 1-17.

[184] Schmidt, A.S. and A.B. Thomsen, *Optimization of wet oxidation pretreatment of wheat straw*. Bioresource Technology, 1998. 64(2): p. 139-151.

[185] Chum, H.L., et al., *Evaluation of pretreatments of biomass for enzymatic hydrolysis of cellulose*. 1985, Solar Energy Research Institute: Golden, Colorado. p. 64

[186] Martín, C., et al., *Wet oxidation pretreatment, enzymatic hydrolysis and simultaneous saccharification and fermentation of clover–ryegrass mixtures*. Bioresource Technology, 2008. 99(18): p. 8777-8782.

[187] Galbe, M. and G. Zacchi, *A review of the production of ethanol from softwood*. Appl Microbiol Biotechnol, 2002. 59: p. 618 - 628.

[188] Vidal, P., *Ozonolysis of Lignin – Improvement of in vitro digestibility of poplar sawdust*. Biomass, 1988. 16(1): p. 1-17.

[189] Energy Information Administration, *Chapter 10 Growth of the Landfill Gas Industry*, in *Renewable Energy Annual 1996*. 1997.DOE/EIA-0603(96)

[190] Breen, A. and F.L. Singleton, *Fungi in lignocellulose breakdown and biopulping*. Current Opinion in Biotechnology, 1999. 10(3): p. 252-258.

[191] Hischier, R., *Life Cycle Inventories of Packaging and Graphical Papers*, in *Ecoinvent Report No.11*. 2007, Swiss Centre for Life Cycle Inventories: Dübendorf

[192] Martínez, Á.T., et al., *Biodegradation of lignocellulosics: microbial, chemical, and enzymatic aspects of the fungal attack of lignin*. International micorbiology, 2005. 8: p. 195-204.

[193] Maijala, P., et al., *Biomechanical pulping of softwood with enzymes and white-rot fungus Physisporinus rivulosus*. Enzyme and Microbial Technology, 2008. 43(2): p. 169-177.

[194] Shukla, O.P., U.N. Rai, and S.V. Subramanyam. *Biopulping and Biobleaching: An Energy and envioronment Saving Technology for Indian Pulp and Paper Industry*. 2004 [cited 2012 April]; Available from: http://isebindia.com/01_04/04-04-3.html.

[195] Ray, M.J., et al., *Brown rot fungal early stage decay mechanism as a biological pretreatment for softwood biomass in biofuel production*. Biomass and Bioenergy, 2010. 34(8): p. 1257-1262.

[196] Wan, C. and Y. Li, *Fungal pretreatment of lignocellulosic biomass*. Biotechnology Advances, (0).

[197] Yu, H., et al., *The effect of biological pretreatment with the selective white-rot fungus Echinodontium taxodii on enzymatic hydrolysis of softwoods and hardwoods*. Bioresource Technology, 2009. 100(21): p. 5170-5175.

[198] Monrroy, M., et al., *Structural change in wood by brown rot fungi and effect on enzymatic hydrolysis*. Enzyme and Microbial Technology, 2011. 49(5): p. 472-477.

Chapter 3

High-Efficiency Separation of Bio-Oil

Shurong Wang

Additional information is available at the end of the chapter

1. Introduction

1.1. What is fast pyrolysis?

Biomass is a CO_2-neutral energy source that has considerable reserve. It can replace fossil feedstock in the production of heat, electricity, transportation fuels, chemicals, and various materials. Liquid bio-fuels, which are considered to be substitutes for traditional petrol liquid fuels, can be produced from biomass in different ways, such as high-pressure liquefaction, hydrothermal pyrolysis, and fast pyrolysis.

Fast pyrolysis is a technology that can efficiently convert biomass feedstock into liquid biofuels. The liquid obtained from fast pyrolysis, which is also called crude bio-oil, may be used as burning oil in boilers or even as a transportation fuel after upgrading. Fast pyrolysis is a process in which lignocellulosic molecules of biomass are rapidly decomposed to short-chain molecules in the absence of oxygen. Under conditions of high heating rate, short residence time, and moderate pyrolysis temperature, pyrolysis vapor and some char are generated. After condensation of the pyrolysis vapor, liquid product can be collected in a yield of up to 70 wt% on a dry weight basis (Bridgwater et al., 1999; Lu et al., 2009). The obvious advantages of the process are as follows:

1. Low-grade biomass feedstock can be transformed into liquid biofuels with relatively higher heating value, thus making storage and transportation more convenient.
2. The by-products are char and gas, which can be used to provide the heat required in the process or be collected for sale.
3. For waste treatment, fast pyrolysis offers a method that can avoid hazards such as heavy metal elements in the char and reduce pollution of the environment.

Many researchers have focused on the techniques of fast pyrolysis, and various configurations of reactor have been developed to satisfy the requirements of high heating rate, moderate reaction temperature, and short vapor residence time for maximizing bio-oil production. During the past decades, many types of reactor have been designed to promote the large-scale and commercial utilization of biomass fast pyrolysis, such as the fluidized bed reactor (Luo et al., 2004; Wang et al., 2002), the ablative reactor (Peacocke & Bridgwater, 1994), the rotating cone reactor (Muggen, 2010; Peacocke; Wagenaar, 1994) and Vacuum reactor (Bridgwater, 1999; Yang et al., 2001).

1.2. The composition and properties of bio-oil

The chemical composition of bio-oil is significantly different from that of petroleum fuels. It consists of different compounds derived from decomposition reactions of cellulose, hemicellulose, and lignin. The chemical composition of bio-oil varies depending on the type of biomass feedstock and the operating parameters. Generally speaking, bio-oil is a mixture of water and complex oxygen-rich organic compounds, including almost all such kinds of organic compounds, that is, alcohols, organic acids, ethers, esters, aldehydes, ketones, phenols, etc. Normally, the component distribution of bio-oil may be measured by GC-MS analysis.

Crude bio-oil derived from lignocellulose is a dark-brown, viscous, yet free-flowing liquid with a pungent odor. Crude bio-oil has an oxygen content of 30–50 wt%, resulting in instability and a low heating value (Oasmaa & C., 2001). The water content of bio-oil ranges from 15 to 50 wt%. The high water content of bio-oil derives from water in the feedstock and dehydration reactions during biomass pyrolysis (Bridgwater, 2012). Heating value is an important indicator for fuel oils. The heating value of bio-oil is usually lower than 20 MJ/kg, much lower than that of fuel oil. The high water content and oxygen content are two factors responsible for its low heating value. The density of bio-oil derived from fast pyrolysis is within the range 1100–1300 kg/m³(Adjaye et al., 1992). The pH value of bio-oil is usually in the range 2–3 owing to the presence of carboxylic acids such as formic acid and acetic acid. The strong acidity can corrode pipework and burner components. Measurements of the corrosiveness of bio-oil have shown that it can induce an apparent mass loss of carbon steel and the breakdown of a diesel engine burner (Wright et al., 2010).

Fresh bio-oil is a homogeneous liquid containing a certain amount of solid particles. After long term storage, it may separate into two layers and heavy components may be deposited at the bottom. As mentioned above, the high content of oxygen and volatile organic compounds are conducive to the ageing problems of bio-oil. The aldol condensation of aldehydes and alcohols and self-aggregation of aldehydes to oligomers are two of the most likely reactions to take place. Coke and inorganic components in the bio-oil may also have a catalytic effect, thereby enhancing the ageing process (Rick & Vix, 1991).

1.3. The utilization of bio-oil

The oxygenated compounds in bio-oil can lead to several problems in its direct combustion, such as instability, low heating value, and high corrosiveness. Although higher water content can improve the flow properties and reduce NOx emissions in the fuel combustion process, it causes many more problems. It not only decreases the heating value of the fuel, but also increases the corrosion of the combustor and can result in flame-out. The low pH value of bio-oil also aggravates corrosiveness problems, which may lead to higher storage and transportation costs. Many researchers have tested the combustion of bio-oil in gas boiler systems, diesel engines, and gas turbines (Czernik & Bridgwater, 2004).

Fresh bio-oil from different feedstocks can generally achieve stable combustion in a boiler system. One problem, however, is the difficulty of ignition. The high water content of bio-oil not only decreases its heating value, but also consumes a large amount of latent heat of vaporization (Bridgwater & Cottam, 1992). Thus, the direct ignition of bio-oil in a cold environment is not easy, and an automatic atomizer assists in ignition and pre-heating of the furnace. The combustion of bio-oil in diesel engines is more challenging. Its long ignition delay time, short burn duration, and lower peak heat release have limited its combustion properties (Vitolo & Ghetti, 1994). Experiments employing bio-oil in gas turbines have proved largely unsuccessful. The high viscosity and high ash content of bio-oil result in severe blocking and attrition problems in the injection system. Moreover, acid in the bio-oil is harmful to the mechanical components of the gas turbine.

Even though many combustion tests of bio-oil have shown its combustion performances to be inferior to those of fossil fuels, the environmental advantages of bio-oil utilization cannot be ignored. Comparative tests have shown that the SO_2 emissions from bio-oil combustion are much lower than those from fossil fuel combustion.

Bio-oil is a mixture of many organic chemicals, such as acetic acid, turpentine, methanol, etc. Many compounds in bio-oil are important chemicals, such as phenols used in the resins industry, volatile organic acids used to produce de-icers, levoglucosan, hydroxyacetaldehyde, and some agents applied in the pharmaceutical, synthetic fiber, and fertilizer industries, as well as flavoring agents for food products (Radlein, 1999). Besides, bio-oil can also be used in a process that converts traditional lime into bio-lime (Dynamotive Corporation, 1995).

2. Separation of bio-oil for upgrading or refinement

2.1. The importance of separation technology

Bio-oil cannot be directly applied as a high-grade fuel because of its inferior properties, such as high water and oxygen contents, acidity, and low heating value. Thus, it is necessary to upgrade bio-oil to produce a high-grade liquid fuel that can be used in engines (Bridgwater, 1996; Czernik & Bridgwater, 2004; Mortensen et al., 2011).

In view of its molecular structure and functional groups, and using existing chemical processes for reference, such as hydrodesulfurization, catalytic cracking, and natural gas steam reforming, several generic bio-oil upgrading technologies have been developed, including hydrogenation, cracking, esterification, emulsification, and steam reforming.

Components with unsaturated bonds, such as aldehydes, ketones, and alkenyl compounds, influence the storage stability of bio-oil, and hydrogenation could be used to improve its overall saturation (Yao et al., 2008). Hydrogenation can achieve a degree of deoxygenation of about 80%, and transform bio-oil into high-quality liquid fuel (Venderbosch et al., 2010; Wildschut et al., 2009). This process requires a high pressure of hydrogen, which increases both the complexity and cost of the operation. Alcohol hydroxyl, carbonyl, and carboxyl groups were easily hydrodeoxygenated, and phenol hydroxyl and ether groups were also reactive, while furans, having a cyclic structure, were more difficult to convert (Furimsky, 2000). After the separation of bio-oil, the components with alcohol hydroxyl, carbonyl, carboxyl, phenol hydroxyl, and ether groups can be efficiently hydrodeoxygenated at a low hydrogen pressure, while the hydrodeoxygenation of more complex components, such as ethers and furans, may be achieved by developing special catalysts.

Catalytic cracking of bio-oil refers to the reaction whereby oxygen is removed in the form of CO, CO_2, and H_2O, in the presence of a solid acid catalyst, such as zeolite, yielding a hydrocarbon-rich high-grade liquid fuel. In the process of cracking, oxygenated compounds in bio-oil are thought to undergo initial deoxygenation to form light olefins, which are then cyclized to form aromatics or undergo some other reactions to produce hydrocarbons (Adjaye & Bakhshi, 1995a). Since bio-oil has a relatively low H/C ratio, and dehydration is accompanied by the loss of hydrogen, the H/C ratio of the final product is generally low, and carbon deposits with large aromatic structures tend to be formed, which can lead to deactivation of the catalyst (Guo et al., 2009a). The cracking of crude bio-oil is always terminated in a short time, with a coke yield of about 20% (Adjaye & Bakhshi, 1995b; Vitolo et al., 1999). Alcohols, ketones, and carboxylic acids are efficiently converted into aromatic hydrocarbons, while aldehydes tend to condense to form carbon deposits (Gayubo et al., 2004b). Phenols also show low reactivity and coking occurs readily (Gayubo et al., 2004a). Besides, some thermally sensitive compounds, such as pyrolitic lignin, might undergo aggregation to form a precipitate, which would block the reactor and lead to deactivation of the catalyst. Consequently, efforts have been made to avoid this phenomenon by separating these compounds through thermal pre-treatment (Valle et al., 2010). Therefore, to maintain the stability and high performance of the cracking process, it is necessary to obtain fractions suitable for cracking by separation of bio-oil, to achieve the partial conversion of bio-oil into hydrocarbon fuels.

Bio-oil has a high content of carboxylic acids, so catalytic esterification is used to neutralize these acids. Both solid acid and base catalysts display high activity for the conversion of carboxylic acids into the corresponding esters, and the heating value of the upgraded oil is thereby increased markedly (Zhang et al., 2006). Since this method is more suitable for the transformation of carboxylic acids, which constitute a relatively small proportion of crude bio-oil, an ester fuel with a high heating value can be expected to be produced from the esterification of a fraction enriched with carboxylic acids obtained from the separation.

The emulsion fuel obtained from bio-oil and diesel is homogeneous and stable, and can be burned in existing engines. Research on the production of emulsions from crude bio-oil and diesel suggested that the emulsion produced was more stable than crude bio-oil. Subsequent tests of these emulsions in different diesel engines showed that because of the presence of carboxylic acids, the injector nozzle was corroded, and this corrosion was accelerated by the high velocity turbulent flow in the spray channels (Chiaramonti et al., 2003a; Chiaramonti et al., 2003b). Besides corrosion, the high water content of bio-oil will lower the heating value of the emulsion as a fuel, and some high molecular weight components such as sugar oligomers and pyrolitic lignin will increase the density and reduce the volatility of the emulsion. Thus, it is beneficial to study the emulsification of the separated fractions that contain less water and fewer high molecular weight components.

Catalytic steam reforming of bio-oil is also an important upgrading technology for converting it into hydrogen. Research on the steam reforming of acetic acid and ethanol is now comparatively mature, with high conversion of reactants, hydrogen yields, and stability of the catalysts (Hu & Lu, 2009). However, some oxygenated compounds in bio-oil show inferior reforming behavior. Phenol cannot be completely converted even at a high steam-to-carbon ratio, while m-cresol and glucose not only show low reactivity, but are also easily coked (Constantinou et al., 2009; Hu & Lu, 2009). To improve the reforming process, some further investigations of steam reforming based on other separating methods are needed.

Therefore, it is necessary to combine crude bio-oil utilization with the current upgrading technologies. Taking advantage of efficient bio-oil separation to achieve the enrichment of compounds in the same family or the components that are suitable for the same upgrading method is a significant strategy for the future utilization of high-grade bio-oil.

2.2. Conventional separation technologies

The efficient separation of bio-oil establishes a solid foundation for its upgrading. Currently, conventional methods for bio-oil separation include column chromatography, solvent extraction, and distillation.

2.2.1. Solvent extraction

The solvents for extraction include water, ethyl acetate, paraffins, ethers, ketones, and alkaline solutions. In recent years, some special solvents, such as supercritical CO_2, have also been used for extraction or other research. By selecting appropriate solvents for extraction of the desired products, good separation of bio-oil can be achieved.

Some researchers have used non-polar solvents for the primary separation of bio-oil, such as toluene and n hexane, and then proceeded to extract the solvent insoluble fraction with water; finally, the water-soluble and water-insoluble fractions were further extracted with diethyl ether and dichloromethane, respectively (Garcia-Perez et al., 2007; Oasmaa et al., 2003). A lot of organic solvents are consumed during the process. Considering the cost of these solvents and the difficulty of the recovery process, the operating costs are unacceptable, which hinders its industrialization.

Supercritical fluid extraction is based on the different dissolving abilities of supercritical solvents under different conditions. Supercritical fluid extraction at low temperatures contributes to preventing undesirable reactions of thermally sensitive components. Researchers usually use CO_2 as the supercritical solvent. In a supercritical CO_2 extraction, compounds of low polarity (aldehydes, ketones, phenols, etc.) are selectively extracted, while acids and water remain in the residue phase (Cui et al., 2010).

2.2.2. Column chromatography

The principle of column chromatography is that substances are separated based on their different adsorption capabilities on a stationary phase. In general, highly polar molecules are easily adsorbed on a stationary phase, while weakly polar molecules are not. Thus, the process of column chromatography involves adsorption, desorption, re-adsorption, and re-desorption. Silica gel is commonly used as the stationary phase, and an eluent is selected according to the polarity of the components. Paraffin eluents, such as hexane and pentane, are used to separate aliphatic compounds. Aromatic compounds are usually eluted with benzene or toluene. Some other polar compounds are obtained by elution with methanol or other polar solvents (Ertas & Alma, 2010; Onay et al., 2006; Putun et al., 1999).

2.2.3. Distillation

Distillation is a common separating technology in the chemical industry. This method separates the components successively according to their different volatilities, and it is essential for the separation of liquid mixtures. Atmospheric pressure distillation, vacuum distillation, steam distillation, and some other types of distillation have been applied in bio-oil separation.

Due to its complex composition, the boiling of bio-oil starts below 100 °C under atmospheric pressure, and then the distillation continues up to 250–280 °C, whereupon 35–50% of residue is left (Czernik & Bridgwater, 2004).

The thermal sensitivity of bio-oil limits the operating temperature of distillation. In view of the unsatisfactory results obtained by atmospheric pressure distillation, researchers have employed vacuum distillation to lower the boiling points of components, and bio-oil could thereby be separated at a low temperature. Characterization of the distilled organic fraction showed that it had a much better quality than the crude bio-oil, containing little water and fewer oxygenated compounds, and having a higher heating value.

Steam distillation is performed by introducing steam into the distilling vessel, to heat the bio-oil and decrease its viscosity, and finally the volatile components are expelled by the steam. In a study combining steam distillation with reduced pressure distillation, bio-oil was first steam distilled to recover 14.9% of a volatile fraction. The recovered fraction was then further distilled by reduced pressure distillation to recover 16 sub-fractions (Murwanashyaka et al., 2001). In this process, a syringol-containing fraction was separated and syringol with a purity of 92.3% was obtained.

Due to its thermal sensitivity, it is difficult to efficiently separate bio-oil by conventional distillation methods. Molecular distillation seems to offer a potential means of realizing bio-oil separation, because it has the advantages of low operating temperature, short heating time, and high separation efficiency.

2.3. Molecular distillation

There are forces between molecules, which can be either repulsive or attractive depending on intermolecular spacing. When molecules are close together, the repulsive force is dominant. When molecules are not very close to each other, the forces acting between them are attractive in nature, and there should be no intermolecular forces if the distance between molecules is very large. Since the distances between gas molecules are large, the intermolecular forces are negligible, except when molecules collide with each other. The distance between collisions with another molecule is called its free path.

The mean free path of an ideal gas molecule can be described by Eq. (1):

$$\lambda_m = \frac{k}{\sqrt{2}\pi \, d^2 p} \, T \tag{1}$$

Where T (°C) is the local temperature; λ_m (m) refers to the mean free path; d (m) is the effective diameter of the molecule; P (Pa) is the local pressure; and k is the Boltzmann constant.

As is apparent from Eq. (1), the molecular mean free path is inversely proportional to the pressure and the square of the effective molecular diameter. Under certain conditions, that is, if the temperature and pressure are fixed, the mean free path is a function of the effective molecular diameter. Apparently, a smaller molecule has a shorter mean free path than a larger molecule. Furthermore, molecular mean free path will increase with increasing temperature or decreasing pressure.

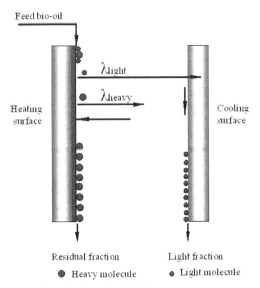

Figure 1. Schematic representation of molecular distillation.

Molecules will move more rapidly when the liquid mixture is heated. Surface molecules will overcome intermolecular forces and escape as gas molecules when they obtain sufficient energy. With an increased amount of gas molecules above the liquid surface, some molecules will return to the surface. Under certain conditions, the molecular motion will achieve dynamic equilibrium, which is manifested as equilibrium on a macroscopic scale.

Traditional distillation technology separates components by differences in their boiling points. However, molecular distillation (or short-path distillation) is quite different and precisely relies on the various mean free paths of different substances. As shown in Fig. 1, the distance between the cooling and heating surfaces is less than the mean free path for a light molecule, but greater than that for a heavy molecule. Therefore, the light molecules escaping from the heating surface can easily reach the cooling surface and be condensed. The dynamic balance is thereby broken, and the light molecules are continuously released from the liquid phase. On the contrary, the heavy molecules are not released and return to the liquid phase. In this way, the light and heavy molecules are effectively separated.

Molecular distillation technology has been widely used in the chemical, pharmaceutical, and foodstuff industries, as well as in scientific research to concentrate and purify organic chemicals. It is a feasible process for the separation of thermally unstable materials, taking into account that it only takes a few seconds to complete the separation process. Bio-oil is a complex mixture of many compounds with a wide range of boiling points. It is thermally sensitive and easily undergoes reactions such as decomposition, polymerization, and oxygenation. Additionally, most of the compounds are present in low concentrations. Molecular distillation is not limited by these unfavorable properties and is suitable for the separation of bio-oil to facilitate analysis and quantification of its constituent compounds.

3. High-efficiency separation of bio-oil at Zhejiang University

3.1. A molecular distillation apparatus

Fig. 2 shows a KDL-5 wiped-film molecular distillation apparatus used for bio-oil separation research at Zhejiang University, which was manufactured by UIC Corporation in Germany. It consists of four main units, namely a feeding unit, an evaporation unit, a condensation unit, and a reduced pressure unit. The feeding unit mainly comprises a graduated dosing funnel with a double jacket, which is filled with heat-transfer oil to control the temperature and to ensure free flowing of the feedstock. The evaporation unit comprises a cylindrical evaporator with a surface area of $0.048\,m^2$, encased in a double jacket containing heat-transfer oil to maintain good temperature homogeneity. It is worth noting that all of the temperatures of these sections are independent. The condensation unit has two cold traps. The first cold trap (or internal condenser) is located in the center of evaporator, and condenses the volatile compounds reaching the cooling surface. There is another cold trap to prevent uncondensed volatile organic compounds from entering the pump. In the reduced pressure unit, the condensation temperature is usually set at $-25\,°C$. The evaporation temperature ranges from room temperature to $250\,°C$, while the operating pressure can be as low as 5 Pa.

Figure 2. KDL-5 molecular distillation apparatus.

The bio-oil used at Zhejiang University was produced from a bench-scale fluidized bed fast pyrolysis reactor (Wang et al., 2008). Crude bio-oil often contains some solid particles, which would abrade the evaporator surface and block the orifice of the dosing funnel, so it is necessary to perform some pre-treatments. Centrifugation and filtration are usually used to remove the solid particles and traditional reduced pressure distillation can also be used to remove water and volatile compounds. The pre-treated bio-oil is placed in the funnel and then the separation process starts. The volatile components released from the thin liquid film are condensed by the internal condenser to form the distilled fraction, while the heavy compounds that are not vaporized flow along the evaporator surface and are collected as the residual fraction.

Because of the short residence time of the feed material at the evaporation temperature, this gentle distillation process only puts a low thermal load on the materials to be distilled. It is therefore appropriate for the separation of bio-oil, which is thermally unstable.

3.2. Single separation process under different operating conditions

3.2.1. Physical characteristics of samples

Bio-oil used in the single separation process was produced by the pyrolysis of Mongolian pine sawdust (Wang et al., 2008). Wang et al. (Guo et al., 2009b; Wang et al., 2009) carried out experimental research on molecular separation of the bio-oil, which was pre-treated by centrifugation and filtration to remove solid particles. Molecular distillation of the bio-oil at

50, 70, 100, and 130 °C, respectively, was investigated under a fixed pressure of 60 Pa. Under all of the tested conditions, the light fraction collected by the second condenser placed before vacuum pump was designated as LF, the middle fraction condensed by the internal condenser as MF, and the heavy fraction as HF.

The color of the distilled fractions becomes lighter while the residual fractions become darker. Under the four conditions, water was concentrated in the LFs, which had water contents of about 70 wt%. The LFs could not be burned because of their high water contents. The pH values of the LFs were in the range 2.13–2.17 as a result of their carboxylic acid contents. On the other hand, the HFs had the highest heating values and the lowest water contents, resulting in good ignitability but inferior fluidity. At a distillation temperature of 70 °C, the water content of the MF was as low as 2 wt%. The total mass of the bio-oil distillation fractions amounted to more than 97% of the bio oil feed. With increasing temperature, the yield of the LF increased without any coking or polymerization problem. Water and volatile carboxylic acids were evaporated from the feedstock in the temperature range 50–130 °C under low pressure, and more carboxylic acids escaped from the liquid at higher temperature. However, on further increasing the temperature, this phenomenon was not so pronounced, due to more and more molecules of higher boiling point also being distilled. The yield of the distilled fraction increased with increasing distillation temperature. However, too high temperature may lead to decomposition of some chemical compounds in the crude bio-oil. Hence, there must be an optimum temperature to realize reasonable separation.

3.2.2. Distribution of acidic compounds in bio-oil fractions

The high content of carboxylic acids in bio-oil is one of the main reasons for its corrosiveness, which damages storage tanks, boilers, and gas turbines. As a consequence, detailed research on the separation of acidic compounds has been carried out under the condition of distillation at 50 °C.

The carboxylic acid content in the refined bio-oil was used to estimate the separation efficiency. Guo et al. (2009b) chose five major acids in bio-oil and studied their separation characteristics. As shown in Fig. 3, the amount of acetic acid, the most abundant acid in bio-oil, was reduced to 1.9 wt% and 0.96 wt% in the MF and HF, respectively. The results indicated that acidic compounds could be effectively separated from the crude bio-oil by means of molecular distillation technology. The LF, which was rich in water and carboxylic acids, was valuable for further catalytic esterification of bio-oil acidic compounds. Both MF and HF could be further upgraded to produce high quality fuels.

3.2.3. Distribution characteristics of several chemicals in three fractions

Fig. 4 illustrates the distributions of selected compounds in bio-oil, MF, and HF. Six chemicals were selected as being representative of ketones, aldehydes, phenols, and sugars, respectively. 1-Hydroxy-2-propanone, the most abundant ketone in bio-oil, could not be detected in the MF or HF after separation, indicating that it was extremely enriched in the

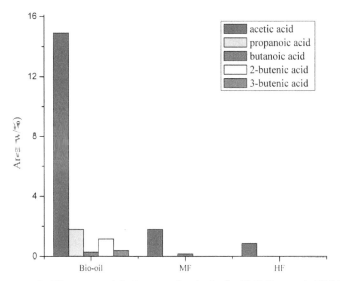

Figure 3. Contents of acidic compounds in three samples obtained at 50 °C (Guo et al., 2009b).

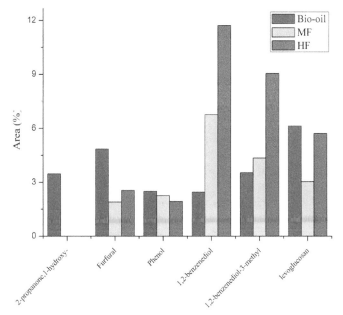

Figure 4. Distributions of selected compounds obtained at 50 °C in three fractions.

LF. The content of furfural in the HF was just a little higher than that in the MF, but much lower than that in bio-oil. The distributions of these two compounds reflected the enrichment of small ketone and aldehyde molecules in the LF. Phenol appeared to be difficult to separate as there were similar distributions in bio-oil, MF, and HF. In contrast, compounds of higher molecular weight tended to be enriched in the MF and HF. For example, 1,2-benzenediol and 3-methyl-1,2-benzenediol were more abundant in the MF and HF than in the bio-oil before separation. In particular, the relative content of 1,2-benzenediol amounted to 11.73 wt% in HF, about five times higher than that in bio-oil (2.45 wt%).

3.2.4. Statistical method to evaluate the separation of bio-oil

As the composition of the bio-oil and the effects of the operating conditions on the distribution of each fraction are both complicated, Wang et al. (2009) put forward a statistical method to directly evaluate the separation level of bio-oil by molecular distillation. The separation coefficients of four groups, "Complete Isolation", "Nonvaporization", "Enrichment", and "Even Distribution", were calculated from the ratios of relative peak of a single component with respect to total components. The results showed that "Complete Isolation" had the largest percentage, followed by "Even Distribution", "Non-vaporization", and "Enrichment" which contained only small parts. Meanwhile, the temperature had a significant effect on the distributions of the compounds.

3.3. Multiple molecular distillation for bio-oil separation

Based on the above single distillation experiments, a multiple molecular distillation experiment was carried out to further evaluate the separation characteristics of bio-oil (Guo et al., 2010b). The feed bio-oil, which was pre-treated by centrifugation, filtration, and vacuum distillation, was firstly distilled at 80 °C and 1600 Pa to obtain the distilled fraction 1 (DF-1) and the residual fraction 1 (RF-1). A part of RF-1 was then further distilled at 340 Pa to obtain DF-2 and RF-2 fractions. In the multiple distillation process, the distilled fraction yield of each distillation process was about 26 wt%. The amounts of water in RF-1 and RF-2 were greatly reduced. The RFs from the two processes had higher heating values than the feed bio-oil or DFs. The acid content was 11.37 wt% in the feed bio-oil, while it was 17.36 wt% for DF-1, nearly four times higher than that in RF-1 (4.56 wt%). In the second process, the acid content of RF-2 was further reduced to 1.38 wt%. The content of monophenols in RF-1 was 36.24 wt%, about twice that in DF-1 (18.02 wt%). Sugars showed non-distillable character in the two distillation processes, and no amounts could be detected in the DF.

In order to gain a deeper insight into the bio-oil distillation properties, Guo (Guo et al., 2010b) proposed a separation factor to evaluate the separation characteristics. The separation factors of acetic acid and 1-hydroxy-2-propanone were approximately 0.9, implying that they could be mostly distilled off. 2-Methoxyphenol, phenol, 2(5H)-furanone, and 2 methoxy-4-methylphenol, the separation factors of which ranged from 0.61 to 0.74, proved to be difficult to separate effectively. Higher molecular weight compounds, such as 3-methoxy-1,2-benzenediol, 4-methoxy-1,2-benzenediol, and 1,2-benzenediol, were very difficult to distil, having separation factors close to zero.

3.4. The joint distillation system at Zhejiang University

Based on the operation experiences gained with the KDL5 molecular distillation apparatus, a larger-scale joint reduced pressure and molecular distillation set-up was established in the State Key Laboratory of Clean Energy Utilization, Zhejiang University. The flow diagram of this joint distillation system is illustrated in Fig. 5. The processing capacities of the reduced pressure distillation and molecular distillation units were both 8–10 kg/h, and they could be run at temperatures up to 300 °C and pressures down to 50 Pa. The reduced pressure distillation unit could be operated separately to remove the water from bio oil as well as to obtain bio-oil fractions. When these two units were assigned to run together, the pre-treated bio-oil from the first reduced pressure distillation unit could be pumped directly into the molecular distillation unit.

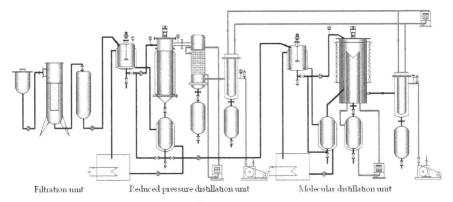

Filtration unit   Reduced pressure distillation unit   Molecular distillation unit

Figure 5. Schematic diagram of the joint distillation system.

3.5. Further research on the distilled fractions

Based on the molecular distillation results, a scheme of the process combining molecular distillation separation with bio-oil upgrading is proposed. The light fraction rich in carboxylic acids and other light components could be used for esterification, catalytic cracking, and steam reforming, to produce ester fuel, hydrocarbons, and hydrogen, respectively. For the middle fraction, steam reforming at high temperature or hydrodeoxygenation at high pressure could efficiently convert this fraction into hydrogen or hydrocarbons. The heavy fraction, which consisted mainly of pyrolytic lignin and sugar oligomers, could be emulsified with diesel to obtain emulsion fuel with a relatively high heating value. On the other hand, the extraction of some valuable chemicals can benefit the overall economy of this process.

Recently, some further research has been performed, aiming at investigating some characteristics of the distilled fractions and devising more promising upgrading methods. Thermal decomposition processes and the pyrolysis products of crude bio-oil and distilled fractions were investigated by means of TG-FTIR by Guo (Guo et al., 2010a). The light fraction (LF) was completely evaporated at 30–150 °C, with the maximum weight loss rate at about 100 °C due to the volatilization of water and compounds of lower boiling point. The middle fraction (MF) and heavy fraction (HF) contained more lignin-derived compounds, and these decomposed continuously over a wide temperature range of 30–600 °C, leaving a final residue yield of 25–30%. Upgrading of the distilled fraction rich in carboxylic acids and ketones was carried out by Guo (Guo et al., 2011). Carboxylic acids accounted for 18.39% of the initial fraction, with acetic acid being the most abundant. After upgrading, the carboxylic acid content decreased to 2.70%, with a conversion yield of 85.3%. The content of esters in the upgraded fraction increased dramatically from 0.72% to 31.1%. The conversion of corrosive carboxylic acids into neutral esters reduced the corrosivity of the bio-oil fraction.

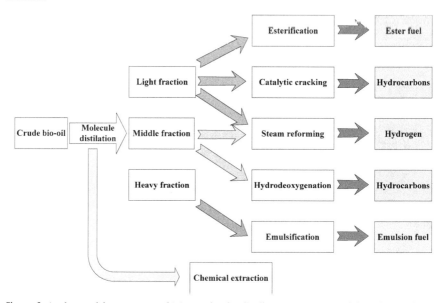

Figure 6. A scheme of the process combining molecular distillation separation with bio-oil upgrading.

Author details

Shurong Wang
Zhejiang University, China

Acknowledgement

The author acknowledges the financial support from the Program for New Century Excellent Talents in University, the International Science & Technology Cooperation Program of China (2009DFA61050), Zhejiang Provincial Natural Science Foundation of China (R1110089), the Research Fund for the Doctoral Program of Higher Education of

China (20090101110034), the National Natural Science Foundation of China (50676085) and the National High Technology Research and Development Program of China (2009AA05Z407). The author also highly appreciates the kind support from Mr. Zuogang Guo, Mr. Qinjie Cai, Mr. Long Guo and Miss Yurong Wang, who have been involved in the experimental research and the preparation of this chapter.

4. References

Adjaye, J.D. & Bakhshi, N.N. (1995a). Catalytic Conversion Of A Biomass-Derived Oil To Fuels And Chemicals .1. Model-Compound Studies And Reaction Pathways. *Biomass & Bioenergy*, Vol.8, No. 3, pp. 131-149, ISSN 0961-9534.

Adjaye, J.D. & Bakhshi, N.N. (1995b). Production Of Hydrocarbons By Catalytic Upgrading Of A Fast Pyrolysis Bio-Oil .1. Conversion Over Various Catalysts. *Fuel Processing Technology*, Vol.45, No.3, pp. 161-183, ISSN 0378-3820.

Adjaye, J.D., Sharma, R.K. & Bakhshi, N.N. (1992). Characterization And Stability Analysis Of Wood-Derived Bio-Oil. *Fuel Processing Technology*, Vol.31, No.3, pp. 241-256, ISSN 0378-3820.

Bridgwater, A.V. (1999). Principles and practice of biomass fast pyrolysis processes for liquids. *Journal of Analytical and Applied Pyrolysis*, Vol.51, No.1-2, pp. 3-22, ISSN 0165-2370.

Bridgwater, A.V. (1996). Production of high grade fuels and chemicals from catalytic pyrolysis of biomass. *Catalysis Today*, Vol.29, No.1-4, pp. 285-295, ISSN 0920-5861.

Bridgwater, A.V. (2012). Review of fast pyrolysis of biomass and product upgrading. *Biomass and Bioenergy*, Vol.38, pp. 68-94, ISSN 0961-9534.

Bridgwater, A.V. & Cottam, M.L. (1992). Opportunities For Biomass Pyrolysis Liquids Production And Upgrading. *Energy & Fuels*, Vol.6, No.2, pp. 113-120, ISSN 0887-0624.

Bridgwater, A.V., Meier, D. & Radlein, D. (1999). An overview of fast pyrolysis of biomass. *Organic Geochemistry*, Vol.30, No.12, pp. 1479-1493, ISSN 1479-1493.

Chiaramonti, D., Bonini, A. & Fratini, E. (2003a). Development of emulsions from biomass pyrolysis liquid and diesel and their use in engines - Part 1: emulsion production. *Biomass & Bioenergy*, Vol.25, No.1, pp. 85-99, ISSN 0961-9534.

Chiaramonti, D., Bonini, A. & Fratini, E. (2003b). Development of emulsions from biomass pyrolysis liquid and diesel and their use in engines - Part 2: tests in diesel engines. *Biomass & Bioenergy*, Vol.25, No.1, pp. 101-111, ISSN 0961-9534.

Constantinou, D.A., Fierro, J.L.G. & Efstathiou, A.M. (2009). The phenol steam reforming reaction towards H_2 production on natural calcite. *Applied Catalysis B-Environmental*, Vol.90, No.3-4, pp. 347-359, ISSN 0926-3373.

Cui, H., Wang, J. & Wei, S. (2010). Supercritical CO2 extraction of bio-oil. *Journal Of Shandong Uniressity Of Technology (Science And Technology)*, Vol.24, No.6, pp. 1-5, 10, ISSN 1672-6197.

Czernik, S. & Bridgwater, A.V. (2004). Overview of applications of biomass fast pyrolysis oil. *Energy & Fuels*, Vol.18, No.2, pp. 590-598, ISSN 0887-0624.

Dynamotive Corporation. (1995). *Acid emission reduction.U.S. Patent 5458803.*

Ertas, M. & Alma, M.H. (2010). Pyrolysis of laurel (Laurus nobilis L.) extraction residues in a fixed-bed reactor: Characterization of bio-oil and bio-char. *Journal of Analytical and Applied Pyrolysis*, Vol.88, No.1, pp. 22-29, ISSN 0165-2370.

Furimsky, E. (2000). Catalytic hydrodeoxygenation. *Applied Catalysis A-General*, Vol.199, No.2, pp. 147-190, ISSN 0926-860X.

Garcia-Perez, M., Chaala, A. & Pakdel, H. (2007). Characterization of bio-oils in chemical families. *Biomass & Bioenergy*, Vol.31, No.4, pp. 222-242, ISSN 0961-9534.

Gayubo, A.G., Aguayo, A.T. & Atutxa, A. (2004a). Transformation of oxygenate components of biomass pyrolysis oil on a HZSM-5 zeolite. I. Alcohols and phenols. *Industrial & Engineering Chemistry Research*, Vol.43, No.11, pp. 2610-2618, ISSN 0888-5885.

Gayubo, A.G., Aguayo, A.T. & Atutxa, A. (2004b). Transformation of oxygenate components of biomass pyrolysis oil on a HZSM-5 zeolite. II. Aldehydes, ketones, and acids. *Industrial & Engineering Chemistry Research*, Vol.43, No.11, pp. 2619-2626, ISSN 0888-5885.

Guo, X.J., Wang, S.R. & Guo, Z.G. (2010a). Pyrolysis characteristics of bio-oil fractions separated by molecular distillation. *Applied Energy*, Vol.87, No.9, pp. 2892-2898, ISSN 0306-2619.

Guo, X.Y., Zheng, Y. & Zhang, B.H. (2009a). Analysis of coke precursor on catalyst and study on regeneration of catalyst in upgrading of bio-oil. *Biomass & Bioenergy*, Vol.33, No.10, pp. 1469-1473, ISSN 0961-9534.

Guo, Z.G., Wang, S.R. & Gu, Y.L. (2010b). Separation characteristics of biomass pyrolysis oil in molecular distillation. *Separation and Purification Technology*, Vol.76, No.1, pp. 52-57, ISSN 1383-5866.

Guo, Z.G., Wang, S.R. & Xu, G.H. (2011). Upgrading Of Bio-Oil Molecular Distillation Fraction With Solid Acid Catalyst. *Bioresources*, Vol.6, No.3, pp. 2539-2550, ISSN 1930-2126.

Guo, Z.G., Wang, S.R. & Zhu, Y.Y. (2009b). Sepatation of acid compounds for refining biomass pyrolysis oil. *Journal of Fuel Chemistry and Technology*, Vol.37, No.1, pp. 49-52, ISSN 1872-5813.

Hu, X. & Lu, G.X. (2007). Investigation of steam reforming of acetic acid to hydrogen over Ni-Co metal catalyst. *Journal of Molecular Catalysis A-Chemical*, Vol.261, No.1, pp. 43-48, ISSN 1381-1169.

Hu, X. & Lu, G.X. (2009). Investigation of the steam reforming of a series of model compounds derived from bio-oil for hydrogen production. *Applied Catalysis B-Environmental*, Vol.88, No.3-4, pp. 376-385, ISSN 0926-3373.

Lu, Q., Li, W.Z. & Zhu, X.F. (2009). Overview of fuel properties of biomass fast pyrolysis oils. *Energy Conversion and Management*, Vol.50, No.5, pp. 1376-1383, ISSN 0196-8904.

Lu, Q., Yang, X.L. & Zhu, X.F. (2008). Analysis on chemical and physical properties of bio-oil pyrolyzed from rice husk. *Journal of Analytical and Applied Pyrolysis*, Vol.82, No.2, pp. 191-198, ISSN 0165-2370.

Luo, Z.Y., Wang, S.R. & Liao, Y.F. (2004). Research on biomass fast pyrolysis for liquid fuel. *Biomass & Bioenergy*, Vol.26, No.5, pp. 455-462, ISSN 0961-9534.

Muggen, G. 2010. Empyro Project Summary. in: *PyNe newsletter*, Vol. 27, pp. 3-5.

Murwanashyaka, J.N., Pakdel, H. & Roy, C. (2001). Seperation of syringol from birch wood-derived vacuum pyrolysis oil. *Separation and Purification Technology*, Vol.24, No.1-2, pp. 155-165, ISSN 1383-5866.

Oasmaa, A., Kuoppala, E. & Solantausta, Y. (2003). Fast pyrolysis of forestry residue. 2. Physicochemical composition of product liquid. *Energy & Fuels*, Vol.17, No.2, pp. 433-443, ISSN 0887-0624.

Onay, O., Gaines, A.F. & Kockar, O.M. (2006). Comparison of the generation of oil by the extraction and the hydropyrolysis of biomass. *Fuel*, Vol.85, No.3, pp. 382-392, ISSN 0016-2361

Peacocke, G.V.C. Techno-economic assessment of power production from the Wellman and BTG fast pyrolysis. *Science in Thermal and Chemical Biomass Conversion*, 2006, Vol.2, pp. 1785-1902

Peacocke, G.V.C. & Bridgwater, A.V. (1994). Ablative Plate Pyrolysis Of Biomass For Liquids. *Biomass & Bioenergy*, Vol.7, No.1-6, pp. 147-154, ISSN 0961-9534.

Putun, A.E., Ozcan, A. & Putun, E. (1999). Pyrolysis of hazelnut shells in a fixed-bed tubular reactor: yields and structural analysis of bio-oil. *Journal of Analytical and Applied Pyrolysis*, Vol.52, No.1, pp. 33-49, ISSN 0165-2370.

Radlein, D. 1999. *The production of Chemicals from Fast Pyrolysis Bio-oils. Fast Pyrolysis of Biomass: A handbook*. CPL Press, Newbury.

Rick, F., Vix, U. 1991. Product standards for pyrolysis products for use as fuel in industrial firing plant. in: *Biomass pyrolysis liquids upgrading and utilization*, Elsevier applied science, pp. 177-218.

Valle, B., Gayubo, A.G. & Aguayo, A.T. (2010). Selective Production of Aromatics by Crude Bio-oil Valorization with a Nickel-Modified HZSM-5 Zeolite Catalyst. *Energy & Fuels*, Vol.24, pp. 2060-2070, ISSN 0887-0624.

Vonderbouch, R.H., Ardiyanti, A.R. & Wildschut, J. (2010). Stabilization of biomass-derived pyrolysis oils. *Journal of Chemical Technology and Biotechnology*, Vol.85, No.5, pp. 674-686, ISSN 0268-2575.

Vitolo, S. & Ghetti, P. (1994). Physical And Combustion Characterization Of Pyrolytic Oils Derived From Biomass Material Upgraded By Catalytic-Hydrogenation. *Fuel*, Vol.73, No.11, pp. 1810-1812, ISSN 0016-2361.

Vitolo, S., Seggiani, M. & Frediani, P. (1999). Catalytic upgrading of pyrolytic oils to fuel over different zeolites. *Fuel*, Vol.78, No.10, pp. 1147-1159, ISSN 0016-2361.

Wagenaar, B.M. 1994. *The rotating cone reactor for rapid thermal solids processing*, PhD, University of Twente.

Wang, Q., Liu, Q. & He, B. (2008). Experimental research on biomass flash pyrolysis for bio-oil in a fluidized bed reactor. *Journal of Engineering Thermophysics*, Vol.29, pp. 885-888, ISSN 0253-231X.

Wang, S.R., Gu, Y.L. & Liu, Q. (2009). Separation of bio-oil by molecular distillation. *Fuel Processing Technology*, Vol.90, No.5, pp. 738-745, ISSN 0378-3820.

Wang, S.R., Luo, Z.Y. & Dong, L.J. (2002). Flash pyrolysis of biomass for bio-oil in a fluidized bed reactor. *Acta Energiae Solaris Sinica*, Vol.1, pp. 185-188, ISSN 0254-0096.

Wildschut, J., Mahfud, F.H. & Venderbosch, R.H. (2009). Hydrotreatment of Fast Pyrolysis Oil Using Heterogeneous Noble-Metal Catalysts. *Industrial & Engineering Chemistry Research*, Vol.48, No. 23, pp. 10324-10334, ISSN 0888-5885.

Wright, M.M., Satrio, J.A., Brown, R.C. 2010. Techno-Economic Analysis of Biomass Fast Pyrolysis to Transportation Fuels. NREL.

Yang, J., Blanchette, D., B, D.C., Roy, C. 2001. *Progress in thermochemical biomass conversion.*

Yao, Y., Wang, S.R. & Luo, Z.Y. (2008). Experimental research on catalytic hydrogenation of light fraction of bio-oil. *Journal of Engineering Thermophysics*, Vol.29, No.4, pp. 715-719, ISSN 0253-231X.

Zhang, Q., Chang, J. & Wang, T.J. (2006). Upgrading bio-oil over different solid catalysts. *Energy & Fuels*, Vol.20, No.6, pp. 2717-2720, ISSN 0887-0624.

Chapter 4

A Real Story of Bioethanol from Biomass: Malaysia Perspective

K.L. Chin and P.S. H'ng

Additional information is available at the end of the chapter

1. Introduction

Rising fossil fuel prices associated with growing demand for energy, and environment concerns are the key factors driving strong interest in renewable energy sources, particular in biofuel. Biofuel refers to any type of fuel whose energy is derived from plant materials. Biofuel which includes solid biomass, liquid fuels and various biogases is among the most rapidly growing renewable energy technologies in recently. Biofuels are commonly divided into two groups based on the technology maturity which using the terms "conventional" and "advanced" for classification. Conventional biofuel technologies include well-established processes that are already producing biofuels on a commercial scale. These biofuels, commonly referred to as first-generation, include sugar- and starch-based ethanol, oil-crop based biodiesel and straight vegetable oil, as well as biogas derived through anaerobic digestion. First generation biofuel processes are useful but limited in most cases: there is a threshold above which they cannot produce enough biofuel without threatening food supplies and biodiversity. Whereas, advanced biofuel technologies are extensions from conventional technologies which some are still in the research and development (R&D), pilot or demonstration phase and they are commonly referred to as second- or third-generation. This category includes hydrotreated vegetable oil (HVO), which is based on animal fat and plant oil, as well as bioethanol based on lignocellulosic biomass, such as cellulosic-ethanol. Although there are wide varieties of advanced biofuels conversion technologies exists today, but they are not commercially available yet. Nevertheless, the most commercializable technology and most used biofuel on the global market is bioethanol.

2. Bioethanol

Bioethanol is chemically known as ethyl alcohol (C_2H_5OH) and produced from fermentation of fermentable sugars (i.e. glucose, sucrose, etc.) from plant sources using micro-organisms (yeasts or bacteria). Bioethanol is a clear colourless liquid, it is biodegradable, low in toxicity and causes little environmental pollution if spilt. In the 1970s, Brazil and the United States (US) started mass production of bioethanol grown from sugarcane and corn respectively. Current interest in bioethanol lies in production derived from lignocellulosic biomass. The most common usage of bioethanol is to power automobiles through mixed with petrol. It can be combined with gasoline in any concentration up to pure ethanol (E100). Anhydrous ethanol, that is, ethanol with at most 1% water, can be blended with gasoline in varying quantities to reduce consumption of petroleum fuels and in attempts reduce air pollution. Bioethanol burns to produce carbon dioxide (CO_2) and water. In addition to that, the use of bioethanol is generally CO_2 neutral. This is achieved because in the growing phase of the plant sources, CO_2 is absorbed by the plant and oxygen is released in the same volume that CO_2 is produced in the combustion of the fuel. This creates an obvious advantage over fossil fuels which only emit CO_2 as well as other poisonous emissions [1].

Blending bioethanol with gasoline help to reduce green house gases (GHG) emissions by oxygenate the fuel mixture so it burns more completely. On a life cycle basis, ethanol produced from corn results in about a 20 percent reduction in GHG emissions relative to gasoline. With improved efficiency and use of renewable energy, this reduction could be as much as 52 percent. In near future, bioethanol produced from cellulose has the potential to cut life cycle GHG emissions by up to 86 percent relative to gasoline as reported in EPA's Emission Facts [2].

3. Bioethanol in use

About 75% of bioethanol produced in the world being used to power automobiles, though it may be used for gasoline additives and other industries such as paints and cosmetics. Ethanol fuel blends are widely sold in the United States, Brazil, Europe and China. The most common blend is 10% ethanol and 90% petrol (E10). Vehicle engines require no modifications to run on E10 and vehicle warranties are unaffected also. However, only flexible fuel vehicles can run on up to 85% ethanol and 15% petrol blends (E85). Since 1976 the Brazilian government has made it mandatory to blend ethanol with gasoline with 5% ethanol and 95% petrol, and in 2007 the legal blend is around 25% ethanol and 75% gasoline (E25). Today, bioethanol contribute around 3% of total road transport fuel globally (on an energy basis) and considerably higher shares are achieved in certain countries [3]. The usage of bioethanol as transport fuel will be even more as the recent European Commission energy roadmap has set a target to increase the use of biofuels for transport from 5.75% from 2010 to 10% by 2020 under the Directive 2003/30/EC.

Bioethanol is also used as primarily gasoline additive and extender due to its high-octane rating. Bioethanol replacing lead as an oxygenate additive for traditional petrols in the form of Ethyl tertiary butyl ether (ETBE). The ethanol is mixed with isobutene (a non-renewable petroleum derivative) to form ETBE. At a 10% mixture, ethanol reduces the likelihood of engine knock, by raising the octane rating.

Beside the usage of bioethanol in fuel industry, bioethanol also can serve a wide range of uses in the pharmaceuticals, cosmetics, beverages and medical sectors as well as for industrial uses. The market potential for bioethanol is therefore not just limited to transport fuel or energy production but has potential to supply the existing chemicals industry. These include for use in acetaldehyde (raw material for other chemicals e.g. binding agent for paints and dyes), acetic acid (raw material for plastics, bleaching agent, preservation), ethylacetate (paints, dyes, plastics, and rubber), detergents, thermol (cold medium for refrigeration units and heat pumps), solvent for spirits industry, cosmetics, print colours and varnish. Isopropyl alcohol (IPA), ethyl acetate (EAC), WABCO antifreeze (disinfectant, cleaning agent for electronic devices, solvents) and vinasse, potassium sulphate (feeding stuffs, fertilizer).

4. Bioethanol technology

Bioethanol can be produced either from conventional or advance biofuel technologies depending on the state of sugars polymerization. The predominant technology for producing bioethanol is through fermentation of sucrose from sugar crops such as sugarcane, sugar beet and sweet sorghum. Bioethanol produced from sugar or starchy materials is categorize under the conventional technology and the bioethanol so called first generation bioethanol. Whereas, at present, much focus is on the bioethanol produced from biomass that possesses lignocellulosic content. This second generation bioethanol or cellulosic ethanol could be produced from abundant low-value material, including wood chips, grasses, crop residues, and municipal waste.

Regardless of the bioethanol technologies used to produce bioethanol, the bioethanol process have to undergo several treatment steps in which normally involves pre-treatment, extraction of fermentable sugars and fermentation. Pre treatment process mainly deals with the preparation of the feedstock into smaller size (higher surface to volume ratio) for ease of sugars extraction. Whereas, extraction process with the aim of transforming the various sugars polymer chains into simple fermentable sugars. Fermentation process is a biological process in which fermentable sugars are converted into cellular energy and thereby produce ethanol and carbon dioxide as metabolic waste products in the absence of oxygen (anaerobic process) using *Saccharomyces cerevisiae*. The theoretical yield of bioethanol is 0.51 g per one gram of glucose consumed during fermentation.

5. Bioethanol conversion yield

Commercial production of bioethanol deals with the biotechnological production from different feedstock. The selection of the most appropriate feedstock for ethanol production strongly depends on the local conditions. Due to the agro-ecological conditions, North American and European countries have based their ethanol industry on the starchy materials. In Brazil, sugarcane is the main feedstock for bioethanol production. World production of ethanol (all grades) in 2010 was nearly 70 billion litres (IEA, 2010). Although many countries produce ethanol from a variety of feedstocks, Brazil and the United States are the major producers of ethanol in the world, each accounting for approximately 35 percent of global production [4].

The theoretical yield of ethanol from sucrose is 163 gallons of ethanol per tonne of sucrose. Factoring in maximum obtainable yield and realistic plant operations, the expected actual recovery would be about 141 gallons per tonne of sucrose [5]. Using [6],[7] and [8] reports, average sugar recovery rates, one tonne of sugarcane would be expected to yield 70 L of ethanol and one tonne of sugar beets would be expected to yield 100 L of ethanol. One tonne of molasses, a byproduct of sugarcane and sugar beet processing, would yield about 260 L of ethanol. Corn had the highest ethanol yield per tonne feedstock (403 L/t), followed by wheat with 350 L/t [9]. A lower ethanol yield per tonne of feedstock was obtained for cassava compared to corn. The ethanol yield from starchy materials were basically higher than sugar containing material because of the higher amount of fermentable sugars (glucose) that may be released from the original starchy material [10].

The conversion of sugar containing material into bioethanol is easier compared to starchy materials and lignocellulosic biomass because previous hydrolysis of the feedstock is not required since this disaccharide can be broken down directly by the yeast cells [11]. Therefore, using raw sugar as a feedstock, one tonne would yield 500 L of ethanol while refined sugar would yield 530 L ethanol. Molasses, from either sugarcane or sugar beets, was found to be the most cost competitive feedstock. The table below summarizes the estimated ethanol production yield and conversion efficiency from starchy and sugar containing materials from all over the world, as well as research ethanol yield produced from lignocellulosic biomasses.

Bioethanol is currently produced from raw materials such as sugar cane, or beet or starch from cereals. Recent interest was on the low cost and abundant availability of lignocellulosic biomass as the potential feedstock for bioethanol production. Lignocellulosic biomass which includes agricultural and forestry residues and waste materials, has the advantage of providing a greater choice of potential feedstock that does not conflict with land-use for food production, and that will be cheaper than conventional bioethanol sources. Many researchers from around the world are now working on transforming lignocellulosic biomass such as straw, and other plant wastes, into "green" gold - cellulosic ethanol. Cellulosic ethanol, a fuel produced from the stalks and stems of plants (rather than only from sugars and starches, as with corn ethanol), is starting to take root in the United States.

The bioconversion of lignocellulosic biomass to monomeric sugars is harder to accomplish than the conversion of starch, presently used for bioethanol production. However, many countries are making efforts to utilize these lignocellulosic biomasses into ethanol; Sweden, Australia, Canada and Japan are planning to invest into lignocellulosic ethanol mill [21]. The highest ethanol yield from lignocellulosic materials was obtained using switchgrass, 201 L/t with 80% conversion efficiency. Ballesteros *et. al* [20] studied on ethanol conversion using woody material such as *Populus nigra* and *Eucalyptus globule* found that the yield of 145 L/t and 137 L/t feedstock and conversion efficiency ranging 59% - 64% was observed. The conversion efficiency for lignocellulosic materials was lower than the conversion efficiency obtained from sugar-containing material and starchy material.

Feedstock		Sugar convertible materials (%)	EtOH yield (L/t)		Conversion efficiency (%)	Source
			Actual ethanol yield	Theoretical ethanol yield		
Sugar containing materials	Sugar cane juice(80% MC)	12	70	78	90	[6]; [7]
	Sugar beet (75% MC)	18	100	116	86	[8]
Starchy materials	Cassava (40% MC)	32	178	207	86	[12]
	Sweet sorghum (14% MC)	15	80	97	82	[13]
	Wheat (14% MC)	66	350	427	82	[14]
	Corn (15% MC)	70	403	452	89	[15]
Lignocellulosic biomass*	Cane bagasse	33	140	213	66	[16]
	Wheat straw	36	140	233	60	[17]
	Corn stalk	35	130	226	63	[18]
	Switchgrass	39	201	252	80	[19]
	Populus nigra	35	151	226	64	[20]
	Eucalyptus globulus	36	138	232	59	[20]
	Brassica carinata	33	128	213	60	[20]

* Note: Sugar convertible materials are referred as cellulose content.

Table 1. Comparative indexes for three main types of bioethanol feedstocks

The selection of the feedstock is in concordance with the interests of each country based on their availability and low cost. Because feedstocks typically account for greater than one-third of the production costs, maximizing the bioethanol yield is imperative [22].

6. Bioethanol from lignocellulosic biomass

Second generation bioethanol which made from lignocellulosic biomass or woody crops, agricultural residues or waste is considered a future replacement for the food crops that are currently used as feedstock for bioethanol production. Technology for producing bioethanol from biomass is moving out of the laboratory and into the commercial place. Breakthroughs in bioethanol technology in the past decade has lead to commercialization of biomass conversion technology. In U.S alone, Six companies were listed by the U.S Environmental

Protection Agency (EPA) as cellulosic ethanol producers, and their combined anticipated production volume is 8 million ethanol-equivalent gallons for coming years [23]. The six companies are DuPont Danisco Cellulosic Ethanol LLC, Fiberight LLC, Fulcrum Bioenergy Inc., Ineos Bio, KL Energy Corp. and ZeaChem Inc. In April 2011, Mossi & Ghisolfi Group (M&G) (Chemtex) commenced construction of a commercial-scale 13 million gallons/year (50 million liters) cellulosic ethanol production facility in Crescentino, Italy. Beside that, there is Abengoa Company, which has a 5m litre/year demonstration plant at Salamanca, Spain. In October 2010, Norway-based cellulosic ethanol technology developer Weyland commenced production at its 200,000 liter (approximately 53,000 gallon) pilot-scale facility in Bergen, Norway. In Asia, Nippon Oil Corporation and other Japanese manufacturers including Toyota Motor Corporation plan to set up a research body to develop cellulose-derived biofuels. The consortium plans to produce 250,000 kilolitres (1.6 million barrels) per year of bioethanol by March 2014. In China, cellulosic ethanol plant engineered by SunOpta Inc. and owned and operated by China Resources Alcohol Corporation that is currently producing cellulosic ethanol from corn stover (stalks and leaves) on a continuous, 24-hour per day basis.

6.1. Process

Various process configurations are possible for the production of bioethanol from lignocellulosic biomass, the most common method for bioethanol conversion technology from lignocellulosic biomass involves three key steps:

Pre-treatment : During biomass pre-treatment lignocellulosic biomass is pre-treated with acids or enzymes in order to reduce the size of the feedstock and to open up the plant structure. Normally, the structure of cellulosic biomass is altered; lignin seal is broken, hemicelluloses is reduced to sugar monomers, and cellulose is made more accessible to the hydrolysis that convert the carbohydrates polymers into fermentable sugars.

Hydrolysis: This is a chemical reaction that releases sugars, which are normally linked together in complex chains. In early biomass conversion processes, acids were used to accomplish this. Recent research has focused on enzyme catalysts called "cellulases" that can attack these chains more efficiently, leading to very high yields of fermentable sugars. Although the decomposition of the material into fermentable sugars is more complicated, the fermentation process step is basically identical for bioethanol from either food crops or lignocellulosic biomass.

Fermentation : Microorganisms that ferment sugars to ethanol include yeasts and bacteria. Research has focused on expanding the range and efficiency of the organisms used to convert sugar to ethanol.

6.1.1. Pre-treatment

The aim of the pretreatment is to break down the lignin structure and disrupt the crystalline structure of cellulose for enhancing acid or enzymes accessibility to the cellulose during hydrolysis step [24],[25]. Lignocellulosic biomass consists of three major components; Cellulose, hemicellulose and lignin and are in the form of highly complex lignocellulosic matrix. Depending on type of lignocellulosic biomass, the lignin content varies from about 10 – 25%, the hemicelluloses content from about 20 – 35% and the cellulose content from about 35 – 50%. Lignin is a polymer of phenyl propanoid units interlinked through a variety of non-hydrolysable C - C and C-O-C bonds. It therefore is a complex molecule with no clear chemical definition as its structure varies with plant species. Hemicellulose is an amorphous heterogenous group of branched polysaccharides. Its structure is characterised by a long linear backbone of one repeating sugar type with short branched side chains composed of acetate and sugars. Cellulose is a linear molecule consisting of repeating cellobiose units held together by Beta- glycosidic linkages. Cellulose is more homogeneous than hemicellulose but is also highly crystalline and highly resistant to depolymerisation. The three components of lignin, hemicellulose and cellulose are tightly bound to each other in the biomass. In fact hemicellulose acts as a bonding agent between cellulose and lignin. In order to convert this biomass to fuel ethanol, the biomass has to be broken up into the individual components first before the molecular chains within each component can be broken up further into simpler molecules.

6.1.2. Hydrolysis

Once the celluloses disconnect from the lignin, acid or enzymes will be used to hydrolyze the newly freed celluloses into simple monosaccharides (mainly glucose). There are three principle methods of extracting sugars from sugars. These are concentrated acid hydrolysis, dilute acid hydrolysis and enzymatic hydrolysis.

6.1.2.1. Concentrated acid hydrolysis process

The primary advantage of the concentrated acid process is the potential for high sugar recovery efficiency [18]. It has been reported that a glucose yield of 72-82% can be achieved from mixed wood chips using such a concentrated acid hydrolysis process [26]. In general, concentrated acid hydrolysis is much more effective than dilute acid hydrolysis [27]. Furthermore, the concentrated-acid processes can operate at low temperature (e.g. 40ºC), which is a clear advantage compared to dilute acid processes. However, the concentration of acid used is very high in this method (e.g. 30-70%), and dilution and heating of the concentrated acid during the hydrolysis process make it extremely corrosive. Therefore, the process requires either expensive alloys or specialized non-metallic constructions, such as ceramic or carbon-brick lining. The acid recovery is an energy-demanding process.

Despite the disadvantages, the concentrated acid process is still of interest. The concentrated acid process offers more potential for cost reductions than the dilute sulfuric acid process [28]. The concentrated acid hydrolysis process works by adding 70-77% sulfuric acid to the pre-treated lignocellulosic biomass. The acid is added in the ratio of 1.25 to 1.5 acid to 1 lignocellulosic biomass and the temperature is controlled at 40-60ºC. Water is then added to dilute the acid to 20-30% and the mixture is again heated to 100ºC for 1 hour. The gel produced from this mixture is then pressed to release an acid sugar mixture. The acid is then recovered partly by anion membranes and partly in the form of H₂S from anaerobic waste water treatment. The process was claimed to have a low overall cost for the ethanol produced [29].

6.1.2.2. Dilute acid hydrolysis

Dilute acid hydrolysis process is similar to the concentrated acid hydrolysis except using very low concentration of sulfuric acid at higher cooking temperature. Biomass is treated with dilute acid at relatively mild conditions which the hemicelluose fraction is hydrolyzed and normally higher temperature is carried out for depolymerisation of cellulose into glucose. The highest yield of hemicellulose derived sugars were found at a temperature of 190°C, and a reaction time of 5 – 10 min, whereas in second stage hydrolysis considerably higher temperature (230 °C) was found for hydrolysis of cellulose [30].

6.1.2.3. Enzymatic hydrolysis

The enzymatic hydrolysis reaction is carried out by means of enzymes that act as catalysts to break the glycosidic bonds. Instead of using acid to hydrolyse the freed cellulose into glucose, enzymes are use to break down the cellulose in a similar way. Bacteria and fungi are the good sources of cellulases, hemicellulases that could be used for the hydrolysis of pretreated lignocellulosics. The enzymatic cocktails are usually mixtures of several hydrolytic enzymes comprising of cellulases, xylanases, hemicellulases and mannanases.

6.1.3. Fermentation process

The hydrolysis process breaks down the cellulostic part of the biomass into glucose solutions that can then be fermented into bioethanol. Yeast *Saccharomyces cerevisiae* is added to the solution, which is then heated at 32°C. The yeast contains an enzyme called zymase, which acts as a catalyst and helps to convert the glucose into bioethanol and carbon dioxide. Fermentation can be performed as a batch, fed batch or continuous process. For batch process, the fermentation process might takes around three days to complete. The choice of most suitable process will depend upon the kinetic properties of microorganisms and type of lignocellulosic hydrolysate in addition to process economics aspects.

The chemical reaction is shown below:

$$C6H12O6 \xrightarrow[Catalyst]{Zymase} 2C2H5OH + 2CO2$$
$$(Glucose) \qquad (Bioethanol) \quad Carbon\ dioxide$$

6.2. Current development in cellulosic bioethanol

At present, much focus is on the development of methods to produce higher recovery yield bioethanol from lignocellulosic biomass. This can be done through two methods; (1) use of pre-treatment to increase the readiness of lignocellulosic biomass for hydrolysis. (2) increase the conversion yield of lignocellulosic biomass into bioethanol through simultaneous fermentation of glucose and xylose into bioethanol.

As mentioned, one barrier to the production of bioethanol from biomass is that the sugars necessary for fermentation are trapped inside the lignocellulosic biomass. Lignocellulosic biomass has evolved to resist degradation and to confer hydrolytic stability and structural robustness to the cell walls of the plants. This robustness is attributable to the crosslinking between the polysaccharides (cellulose and hemicellulose) and the lignin via ester and ether linkages. Ester linkages arise between oxidized sugars, the uronic acids, and the phenols and phenylpropanols functionalities of the lignin. The cellulose fraction can be only hydrolysed to glucose after a pre-treatment aiming at hydrolytic cleavage of its partially crystalline structure. A number of pre-treatment methods are now available – steam explosion, dilute acid pre-treatment [31] and hydrothermal treatment [32]. Hydrothermal treatment prevent the degradation of cellulose content inside the lignocellulosic biomass during pre-treatment because hydrothermal can be performed without addition of chemicals and oxygen to the lignocellulosic biomass. Hydrothermal treatment involves two process where during the first process, lignocelluosic biomass was soaked in water at 80 °C to soften it before being treated in the second process with higher temperature at 190–200°C.

Another way to increase the recovery yield of bioethanol from lignocellulosic biomass is to convert every bit of biomass into bioethanol. This means using all the available sugars from cellulose and hemicelluose and fermented into bioethanol. Lignocellulosic biomass have high percentage of pentoses in the hemicellulose, such as xylose, or wood sugar, arabinose, mannose, glucose and galactose with majority sugar in hemicelluloses is xylose which account more than 90% present. Unlike glucose, xylose is difficult to ferment. This meant that as much as 25% of the sugars in biomass were out of bounds as far as ethanol production was concerned. At the moment, research shows that steam explosion or mild acid treatment performed under adequate temperature and time of incubation, render soluble the biomass hemicellulose part with the formation of oligomers and C5 sugars that are easily extracted from the biomass. The C5 sugar stream can be individually fermented to ethanol by microorganisms such as *E.coli*, *Pichia stipitis* and *Pachysolen* , that are able to metabolise xylose, or be used as carbon source in a variety of other fermentative processes [33].

7. Bioethanol from lignocellulosic biomass - Malaysia perspective

Malaysia formulated the National Biofuel Policy with envisions to put the biofuel as one of the five energy sources for Malaysia, enhancing the nation's prosperity and well being. This is in line with nation's Five-Fuel Diversification Policy, a national policy to promote renewable energy (RE) as the fifth fuel along with fossil fuels and hydropower. The National Biofuel Policy was implemented in March 2006 to encourage the production of Biofuels, particularly biodiesel from palm oil, for local use and for export. However, in 2007, the Government has announced that the implementation of the whole biodiesel project has been put on hold indefinitely owing to the current high price of refined, bleached and deodorized palm olein.

Recently, the Government of Malaysia launched new strategy to promote the biofuel through the National Biomass Strategy 2020 on year 2011. The aim of National Biomass

Strategy 2020 is to create higher value-added biomass economic activities that contribute towards Malaysia's gross national income (GNI) and creating high value jobs for the benefit of Malaysians. This Strategy outline the production of bioethanol produced from lignocellulosic biomass particularly the oil palm biomass as a starting point with extended to include biomass from other sources such as wood waste. The palm oil sector correspondingly generates the largest amount of biomass, around 80 million dry tonnes in 2010. This is expected to increase to about 100 million dry tonnes by 2020, primarily driven by increases in plantation area. A conservative estimation of utilising an addition 20 million tonnes of oil palm biomass for bioethanol has the potential to contribute significantly to the nation's economy while at the same time reduce the green house gasses emission.

The National Biomass Strategy 2020 proposes a mandate of bioethanol blending of 10 percent in petrol fuel in Malaysia by 2020 to cut down the green house gasses emissions. This would generate a domestic demand for one million tonnes of bioethanol per annum with the first bioethanol from lignocellulosic biomass plant is expected to be commercially viable between 2013 and 2015 [34]. As a result, much attention has been focuses on generating bioethanol from oil palm biomass and wood waste.

As mentioned early, bioethanol utilization as automobile fuel is especially promising as the United States, Brazil and Europe has introduced. However, low-cost supply associated with high bioethanol yield of the bioethanol is indispensable for its wide use. The discussion of economic feasibility of bioethanol production from lignocellulosic biomass in Malaysia in this paper was based on the experimental data through laboratory worked done by [35] and [36] and comparison was made with sugarcane and corn.

7.1. Experiment data

Optimum cellulose conversion to glucose with the hydrolysis efficiency of 82%, 67% and 66% for oil palm trunk, rubberwood and mixed hardwood, respectively obtained using two-stage concentrated sulfuric acid hydrolysis at elevated temperature using 60% sulfuric acid treated in a water bath with a temperature of 60°C for 30 min at the first stage hydrolysis and subsequently subjected to 30% sulfuric acid at 80°C for 60 min at the second stage [36]. As stated in the study by [35], optimum fermentation parameters for lignocellulosic hydrolysates using *Saccharomyces cerevisae* was obtained using 33.2°C and pH 5.3 with the fermentation efficiency of 80%, 85% and 90% for oil palm trunk, rubberwood and mixed hardwood, respectively. The optimum cellulose conversion and fermentation efficiency were used to calculate the actual ethanol yield per tonne (L/t) and the conversion efficiency of lignocellulosic biomass. The conversion efficiency was calculated in percentage of actual yield over the theoretical yield. The theoretical yield was calculated in assumptions that all the cellulose was converted to glucose and further converted to ethanol theoretical yield (51%) in 100% conversion rate. Using the cellulose conversion and fermentation efficiencies, the actual ethanol yields per tonne lignocellulosic biomass can be calculated for lignocellulosic biomass as the equation below:

Ethanol yield in   [1000 (kg) x cellulose content x actual hydrolysis efficiency

liter per tonne of   =   x ethanol theoretical yield (0.51) x actual fermentation

feedstock (L/t)   efficiency] / 0.789

(Note: Ethanol has a density of 0.789 kg/L)

The results of bioethanol yield per tonne for oil palm trunk, rubberwood and mixed hardwood and their conversion efficiencies were presented in Table 2.

	Oil palm trunk	Rubberwood	Mixed hardwood
Celulose content	0.48	0.56	0.56
Hydrolysis efficiency	0.82	0.67	0.66
Ethanol theoretical yield at 100% fermentation efficiency	0.51	0.51	0.51
Actual fermentation efficiency	0.80	0.85	0.90
Actual Ethanol Yield/tonne of dried raw materials	204 L	206 L	215 L
Theoretical Ethanol Yield/tonne of dried raw materials	310 L	362 L	361 L
Total Ethanol Conversion efficiency	66%	57%	60%

Table 2. Ethanol Yield Per Tonne of Feedstock And The Ethanol Conversion Efficiency

As shown from the Table 2, using the same amount of feedstock, mixed hardwood produced slightly higher in volume of bioethanol (215 L/t) compared to oil palm trunk and rubberwood with the ethanol yield per tonne of 204 L/t and 206L/t, respectively. The volume of bioethanol produced using oil palm trunk, rubberwood and mixed hardwood per metric tones of dry weight basically were higher than those reported by [20] as shown in Table 1. The highest conversion efficiency was obtained from oil palm trunk (66%), followed by mixed hardwood (60%) and rubberwood (57%).

If bioethanol yield per tonne feedstock values are taken into consideration, the three lignocellulosic biomass studied was higher than most of the comparing feedstock. The three lignocellulosic biomass ethanol yields per tonne of feedstock were much higher than sugarcane, sugarbeet and cassava. This could be explain by the high moisture content of the sugarcane, sugarbeet and cassava implies the use of a greater amount of feedstock to reach the same sugar content that may released from the lignocellulosic material. However, lower bioethanol yield per tonne feedstock of the studied lignocellulosic biomass was found to be lowered than those wheat and corn feedstocks. This is due to the higher glucose convertible substance in the wheat and corn which contributed to higher ethanol yield. Overall, the conversion efficiency for the studied lignocellulosic biomass was lower than sugar containing material and starchy material. This showed how critical the hydrolysis and fermentation efficiency of the lignocellulosic biomass contributed to a higher ethanol yield to make it comparative with these commercial feedstocks. The three lignocellulosic biomass used in this study in terms of ethanol yield per tonne feedstock were found to be comparable with the results obtained from the lignocellulosic biomass obtained from other studies and conversion efficiency (Table 1). The studied lignocellulosic biomass contained higher amount of cellulose as the glucose convertible material. Therefore, this may contributed to higher ethanol yield per tonne of feedstock.

7.2. Economic feasibility of bioethanol

The cost of biethanol per litre presented here mainly calculated from the cost of raw materials used; i.e. lignocellulosic biomass and sulfuric acid and processing cost. Fixed operating costs are excluded from this calculation. Fixed operating costs including labour and various overhead items are fully incurred regardless of the operating production capacity and their contribution to the total cost of bioethanol is estimated at 15 to 18%. [37] stated that cost of biomass contribute almost 60% to the total production cost which is the highest contributor to the cost of bioethanol. Therefore, the main focus here is to estimate the effect of raw materials price on the cost of bioethanol.

7.2.1. Cost of lignocellulosic biomass

Assessing the various costs of mobilising lignocellulosic biomass today which include harvesting, collection, pre-processing, substitution and transportation to a downstream hub, the order of biomass can be mobilised at globally competitive costs, i.e., at a cost of less than RM 250 per dry-weight tonne. The distance of transportation should be less than 100km in radius from the collection area.

7.2.2. Cost of sulfuric acid and recovery charge

The sulfuric acid is sells at RM 264 per tonne. By far, sulfuric acid is the largest expenditure of raw materials in the process of making bioethanol from lignocellulosic biomass. Nonetheless, the current technology enable the acid-sugar solution from hydrolysis separated into acid and sugar components by means of chromatographic separation using commercial available ion exchange resins to separate the components without diluting the sugar. The separated sulfuric acid is recirculated and reconcentrated to the level required by the decrystallization and hydrolysis steps. Using this technology almost up to 100% of the sulfuric acid can be recovered from the process.

7.2.3. State of art scenario

The state of art scenario presented here makes use of the conversion rates from the experiment data (Table 2). Approximately, 200 L of bioethanol yields per dry tones of lignocellulosic biomass and anticipated prices of RM 250 per dry tones of lignocellulosic biomass and RM 261 per tones of 60% concentrated sulfuric acid. The feedstock cost for one litre of bioethanol produced using either from oil palm trunk or wood wastes is estimated at about RM 1.25/litre. The production cost for one litre bioethanol from lignocellulosic biomass is estimated at RM 0.26 with the hydrolysis cost contributed RM 0.20 based on the sulfuric acid is added at a ratio of 5:1 (acid: dry weight of biomass) with acid lost in the sugar stream is not more than 3% during recovery (97% recoverable). Fermentation cost contributed RM 0.06 with the yeast would be grown at the site without cultivation process [38]. Therefore, the total cost per litre of bioethanol produced is RM 1.51 excluding capital and fixed variable costs. However, without the recovery of sulfuric acid during hydrolysis, the cost of bioethanol from lignocellulosic biomass would be rose up to RM7.85, excluding capital and fixed variable costs. With ethanol prices now at RM 2.10 per litre, it is possible for the Malaysia to produce the bioethanol from oil palm trunk and wood wastes, yet it would be not profitable to produce ethanol from lignocellulosic biomass without using the recovery system for sulfuric acid during hydrolysis.

The table below shows different scenario on the biomass feedstock and bioethanol yield that might affect the cost of bioethanol in Malaysia. The scenarios were based on 97% sulfuric acid recovered during hydrolysis and no change on the cost of fermentation production.

Scenario Analysis :

The economic feasibility of bioethanol production in Malaysia from lignocellulosic biomass is highly dependent on the feedstock cost and recovery yield. The cost of feedstock contributed approximately 80% (excluding capital and fixed variable costs) to the total bioethanol cost when the feedstock price estimated at RM 250 per dry weight ton. As the feedstock price increase 5% to 15% per dry ton, the cost of bioethanol increased from as low as 4% up to almost 13%. Higher recovery yield from the bioethanol process will surely reduce the cost of bioethanol produced per litre when the cost of feedstock remains the same. However, as the conversion yield of bioethanol decrease from 200 L per dry weight ton of biomass, the cost of biothenol per litre increase from 5% up to 17%.

Like corn in the United States and sugarcane in Brazil, the relatively low feedstock cost will only makes this process economically competitive. The cost of producing ethanol from sugarcane in Brazil is estimated at about RM 0.60 per litre, excluding capital costs. U.S. ethanol conversion rates utilizing corn as the feedstock are estimated at approximately 2.65

gallons of ethanol per bushel for a wet mill process and 2.75 gallons per bushel for a dry mill process. Net feedstock costs for a wet mill plant are estimated at about RM 0.30 per litre with total ethanol production costs estimated at RM 0.76 per litre. Net feedstock costs for a dry mill plant are estimated at RM 0.38 per litre with total ethanol production costs at RM 0.76 per litre. Molasses, from either sugarcane or sugar beets, was found to be the most cost competitive feedstock beside the lignocellulosic biomass. Estimated ethanol production costs using molasses were approximately RM 0.92 per litre with a RM 0.66 per litre feedstock cost [39].

	Bioethanol yield (L/T)	Feedstock Price per ton (RM)	Price of Sulfuric Acid per ton (RM)	Cost of Feedstock per litre of bioethanol (RM)	Cost of Production per litre (RM)	Cost of bioethanol per litre (RM)
Laboratory worked	200	250.00	264.00	1.25	0.26	1.51
Scenario 1: Reducing in conversion yield	-5%	Remain	Remain	1.31	0.27	1.58 (+4.6%)
	-10%	Remain	Remain	1.39	0.28	1.67 (+10.6%)
	-15%	Remain	Remain	1.47	0.29	1.76 (+16.6%)
Scenario 2: Increase of feedstock cost	Remain	+5%	Remain	1.31	Remain	1.57 (4.0%)
	Remain	+10%	Remain	1.38	Remain	1.64 (+10.6%)
	Remain	+15%	Remain	1.44	Remain	1.70 (+12.6%)
Scenario 3: Increase of sulfuric acid cost	Remain	Remain	+5%	Remain	0.27	1.52 (+0.6%)
	Remain	Remain	+10%	Remain	0.28	1.53 (+1.3%)
	Remain	Remain	+15%	Remain	0.29	1.54 (+2.0%)

Table 3. Cost of bioethanol per litre with different scenario on cost of raw materials and conversion yield

8. Conclusion

The studied lignocellulosic biomass has a higher bioethanol yield per tonne feedstock (L/t) than most of the commercialized bioethanol feedstock. However, improvement had to be made on the conversion efficiency to obtained higher ethanol yield to make it more comparable with the sugar containing and starchy material. The composition of substance that can be converted to glucose played a big influence on the ethanol yield per tonne feedstock. With the large amount of glucose convertible material and abundant availability, these lignocellulosic biomasses are potential feedstock for bioethanol production.

Author details

K.L. Chin[1] and P.S. H'ng
Faculty of Forestry, Universiti Putra Malaysia, UPM Serdang, Selangor, Malaysia

9. References

[1] Wang M, Saricks C, and Santini D (1999) Effects of Fuel Ethanol Use on Fuel-Cycle Energy and Greenhouse Gas Emissions. Argonne, IL.: USDOE Argonne National laboratory, Center for Transportation Research.

[2] EPA (2007) Emission Facts; Greenhouse Gas Impacts of Expanded Renewable and Alternative Fuels Use. Emission Facts Report (EPA420-F-07-035). Office of Transportation and Air Quality, EPA. US.

[3] IEA (2010) Medium Term Oil and Gas Markets 2010, OECD/IEA, Paris.

[4] F.O. Lights (2011) World Ethanol and Biofuels Report, F.O. Lights, Vol. 10, No. 8.

[5] Bolling C, Suarez NR (2001) The Brazilian Sugar Industry: Recent Developments. Sugar and Sweetener Situation and Outlook. U.S. Department of Agriculture Economic Research Service, 222-232, Septemper 2001.

[6] Baucum LE, Rice RW, Schuneman TG (2006) ,. Electronic Data Information Source (EDIS) SC032. Agronomy Department, University of Florida, Gainesville, FL.

[7] Coelho S (2005) Brazilian Sugarcane Ethanol: Lessons Learned. The Brazilian Reference Center on Biomass. STAP Workshop on Liquid Biofuel, New Delhi, India, 29 August-1 September, 2005.

[8] Berg C (2001) World Fuel Ethanol. Analysis and Outlook. F.O. Lichts.

[9] Pongsawatmanit R, Temsiripong T, Suwonsichon T (2007) Thermal And Rheological Properties Of Tapioca Starch And Xyloglucan Mixtures In The Presence Of Sucrose. Food research international. 40: 239–248 June.

[1] Corresponding Author

[10] Yoosin S, Sorapipatana C (2007) A Study Of Ethanol Production Cost For Gasoline Substitution In Thailand And Its Competitiveness. Thammasat int. j. sc. tech. 12:69-80.

[11] Cardona CA, Sanchez OJ (2007) Fuel Ethanol Production: Process Design Trends and Integration Opportunities. Bioresour. technol. 98: 2415-2457.

[12] Wang W (2002) Cassava Production For Industrial Utilization In China - Present And Future Perspective. In: Cassava research and development in Asia: exploring new opportunities for an ancient crop. Seventh regional cassava workshop. Bangkok, Thailand, October 28–November 1, 2002. pp.33–38.

[13] Rao DB, Ratnavathi CV, Karthikeyan K, Biswas PK, Rao SS, Vijay Kumar BS, Seetharama N (2004) Sweet Sorghum Cane For Bio-Fuel Production: A SWOT Analysis In Indian Context. National Research Centre for Sorghum. pp. 20.

[14] Agu R, Bringhurst T, Brosnan J (2006) Production of grain whisky and ethanol from wheat, maize and other cereals. Journal of the Institute of Brewing. 112(4): 314-323.

[15] Baker A., Zahniser S. (2006) Ethanol Reshapes the Corn Market. Amber Waves. Economic Research Service/USDA. 4(2): pp 30-35. Available: http://www.bioenergypro.com/bioenergy-articles/ethanol/ethanol-reshapes-the-corn-market. Accessed 2012 March 20.

[16] Moreira JS (2000) Sugarcane for Energy – Recent Results and Progress In Brazil. Energy for sustainable development. 6(3): 43–54.

[17] Ballesteros I, Negro MJ, Oliva JM, Cabañas A, Manzanares P, Ballesteros (2006) Ethanol production from steam-explosion pretreated wheat straw. Appl Biochem Biotechnol.129-132:496-508.

[18] Demirbas A (2005) Bioethanol From Cellulosic Materials: A Renewable Motor Fuel From Biomass. Energy sources. 27: 327–337.

[19] Bakker RR, Gosselink RJA, Maas RHW, de Vrije T, de Jong E, van Groenestijn JW, Hazewinkel JHO (2004) Biofuel Production From Acid-Impregnated Willow And Switchgrass In 2nd World Conference on Biomass for Energy Industry and Climate Protection. 10–14 May 2004, Rome, Italy. pp. 1467–1470.

[20] Ballesteros M, Oliva JM, Negro MJ, Manzanares PI (2004) Ethanol From Lignocellulosic Materials By A Simultaneous Saccharification And Fermentation Process (SFS) With *Kluyveromyces marxianus* CECT 10875. Process Biochemistry. 39(12):1843-1848.

[21] F.O. Lights (2008) World Ethanol and Biofuel Report. F.O. Lights. Vol.4, No.7. pp.71-92.

[22] Sánchez OJ, Cardona CA (2008) Trends In Biotechnological Production Of Fuel Ethanol From Different Feedstocks. Bioresour. technol. 99: 5270-5295.

[23] Bevill K (2001) Ethanol Producer Magazine, BBI International.

[24] Mosier N, Wyman CE, Dale BD, Elander RT, Lee YY, Holtzapple M, Ladisch CM (2005) Features Of Promising Technologies For Pretreatment Of Lignocellulosic Biomass. Bioresour. technol. 96: 673–686.

[25] Puls J, Schuseil J (1993) Chemistry Of Hemicelluloses: Relationship Between Hemicellulose Structure And Enzymes Required For Hydrolysis. In: M.P. Coughlan, G.P. Hazlewood, editors. Hemicellulose and Hemicellulases, Portland Press Research Monograph. pp. 1–27.

[26] Iranmahboob J, Nadim F, Moemi S (2002) Optimizing Acid-Hydrolysis: A Critical Step For Production Of Ethanol From Mixed Wood Chips. Biomass and bioenergy. 22: 401–404.

[27] Grahmann K, Torget R, Himmel M (1985) Optimization Of Dilute Acid Pretreatment Of Biomass. Biotechnology and bioengineering symposium. 15: 59–80.

[28] Farooqi R, Sam AG (2004) Ethanol as A Transportation Fuel. Centre for Applied Business Research in Energy and the Environment (CARBREE) Climate Change Inititive. University of Alberta, Canada.

[29] Van Groenestijn J, Hazewinkel O, Bakker R (2006) Pretreatment Of Lignocellulose With Biological Acid Recycling (Biosulfurol Process). Zuckerindustrie. 131(9): 639-641.

[30] Chandel AK, Chan ES, Rudravaram R, Narasu ML, Rao LV, Ravindra P (2007) Economics and Environmental Impact of Bioethanol Production Technologies: An Appraisal. Biotecnology and molecular biology review. 2(1): 14-32.

[31] Schell DJ, Farmer J, Newman M and McMillan JD (2003) Dilute-Sulfuric Acid Pretreatment of Corn Stover in Pilot-Scale Reactor: Investigation of Yields, Kinetics, and Enzymatic Digestibilities of Solids. Applied biochemistry and biotechnology. 105-108: 69-86.

[32] Thomsen MH, Thygesen A, Jorgensen H, Larsen J, Christensen BH, Thomsen AB (2006) Preliminary Results On Optimizing Of Pilot Scale Pretreatment Of Wheat Straw Used In Coproduction Of Bioethanol And Electricity. Applied Biochemistry and Biotechnology. 129–132. 448–460.

[33] Bon EPS, Ferrara MA (2007) The Role of Agricultural Biotechnologies for Production of Bioenergy in Developing Countries. FAO seminar, Rome, 12 October 2007.

[34] AIM, 2011. National Biomass Strategy 2020: New wealth creation for Malaysia's palm oil industry. Agensi Inovasi Malaysia. Putrajaya.

[35] Chin KL, H'ng PS, Wong LJ, Tey BT, Paridah MT (2010) Optimization Study Of Ethanolic Fermentation From Oil Palm Trunk, Rubberwood and Mixed Hardwood Hydrolysates Using Saccharomyces cerevisiae. Bioresour. technol. 101. 3287-3291.

[36] Chin KL, H'ng PS, Wong LJ, Tey BT, Paridah MT (2011) Production Of Glucose From Oil Palm Trunk and Sawdust Of Rubberwood and Mixed Hardwood. Applied energy. 88: 4222-4228.

[37] Wang L, Sharifzadeh M, Templer R, Murphy RJ (2012) Technology Performance And Economic Feasibility Of Bioethanol Production From Various Waste Papers. Energy environ. sci. 5: 5717–5730.

[38] Farone WA, Cuzens JE (1996) Method of Separating Acids and sugars Resulting from Strong Acid Hydrolysis. U.S. Patent No. 5,580,389. December 3.

[39] Shapouri H, Salassi M, Fairbanks JN (2006) The Economic Feasibility Of Ethanol Production From Sugar In The United States. USDA's Renewable Energy. United State Department of Agriculture, Office of the Chief Economist, Office of Energy Policy and New Uses.

Biomass Production

Chapter 5

Biomass from the Sea

Ernesto A. Chávez and Alejandra Chávez-Hidalgo

Additional information is available at the end of the chapter

1. Introduction

In the world oceans there is large amount of biomass suspended in the photic zone of water column. Part of the living part is of plant origin, the phytoplankton and other is the animal component or zooplankton. There is also large proportion of particulate organic matter composed by remains of dead animals and feces. They represent the basis of the food webs with three or four trophic levels where all the consumers are animals in whose top the carnivores or top predators, are found. In all aquatic trophic webs, many species are exploited.

2. The primary producers

Nearly 0.3% of solar energy incident on the sea surface is fixed by phytoplankton the tiny plant organisms suspended in natural waters, over the 40-60 meters of the upper water column, accounts to 75% of primary productivity of an area of the word oceans near to $3.5 \cdot 10^{14}$ square meters; the remaining 25% is produced by macro algae. The amount of biomass of all the consumers is based upon primary production by phytoplankton, which range between $0.05 - 0.5$ gCm^{-2}d^{-1}, but in some very productive upwelling zones or in some grass beds, it can be as high as 5 gCm^{-2}d^{-1} (Russell-Hunter 1970; Margalef 1974; Cushing and Walsh 1976). As a result of this photosynthetic process, the carbon gross production of the sea amounts to 15.5×10^{10} mt of Carbon per year, equivalent to a net production of 1.5×10^{10} mt, most of it in shore waters. By having in mind the energetic efficiency, these figures amount to 8 per cent of global aquatic primary production (Pauly and Christensen 1995; Friedland et al. 2012), meaning that there is a maximum limit to fisheries production.

Biological production through the fixation of light is a process interacting with the degradation or dissipation of energy by all organisms; in other words, the persistence of life as we know it, depends on a permanent input of energy, which after being fixed and transformed in chemical energy by the plants, is dissipated constantly by all organisms on Earth. Human beings have been able to simplify the food webs channelizing the production of a few species which are exploited by man; agriculture systems are a typical example of this process. However, this implies a limit to the maximum potential production of biomass by organisms (Pauly and Christensen 1995; Friedland et al. 2012; Botsford 2012).

3. The secondary producers and food webs

The thermodynamics of the biomass flow and secondary production indicates that the transfer efficiency of carbon in the sea webs may approach to 15 per cent; however, many other authors (in Christensen and Pauly 1993; Pauly and Christensen 1995) adopted the value of 10 per cent. All of the consumers depend from the chemical energy to subsist; this energy is synthesized by the primary producers and transferred to other trophic levels trough consumption by herbivores and then passed to several levels of animals through predation.

Zooplankton, the free-living animals suspended at the water column, are the kind or organisms which make use of the primary production. The main component of this food webs is the group of copepods. Apart o being composed mostly by herbivores, zooplankton also contains many predators of first order, like jelly fish and other crustaceans as larval stages of benthic organisms spending in most cases, from a few days to several months suspended in the water column as predators of micro zooplankton, then being recruited to the benthic communities as they grow.

Caloric value of organisms indicates very uniform qualities through the food web, being higher in those animals storing lipids in their bodies. In sugars and proteins, the caloric value is 4,100 cal g^{-1}, whilst in lipids this value amounts to 9,300 cal g^{-1}, but when these substances are not totally oxidized, the calories available are nearly 90% of their total caloric values. A high production of biomass from the primary producers would be uptaken by the herbivores and transferred to upper levels of the food web. This means that a high primary production will imply high biomass of consumers in proportion following the rule of 10 per cent; this is, for each ton of top predators, there will be 10 mt of predators of first order, and 100 mt of herbivores. The biomass of the carnivores ranges between 0.5 and 2 g C m^{-2} and follows the 10% rule respecting to the lower level. The biomass of primary producers, mainly phytoplankton, may be lower than the herbivores because of their high turnover rate. It is pertinent to mention that upwelling zones of the sea, like in Peru on the west coast of America and West Africa, significant amounts of nutrients are flowing up from the deep sea enriching the surface waters in the photic zone and stimulating the primary productivity. In these zones, the process of evolution has allowed the organization of short food chains, where the sardine and anchovies take advantage exploiting much of this production, allowing the growth of large schools which are exploited by human beings, with levels of exploitation of more than 12 Million mt, as occurred in Peru in the early seventies.

4. The fisheries

The exploitation of aquatic populations by human beings through fisheries, leads to a change in the trophic structure of ecosystems, allowing that opportunistic species, formerly infrequent, to become abundant and reducing the biodiversity; this seems to be the case of squids and jellyfishes. This process determines an increase of the primary production/biomass ratio in the ecosystem. The most productive ecosystems are those associated to upwelling, where the fast growing predators with short life spans, plankton feeders determining the existence of short food chains, allow the existence of very productive fisheries as in the case of anchovy and sardine fisheries. In other natural communities, where the ecosystem usually imposes high environmental stability, top predators usually are animals with long life span in relatively long food chains; in this case, the potential biomass production is low, because the evolutionary forces are oriented towards the density dependent processes, leading to the organization of ecosystems with high biodiversity as occurs in coral reefs. In this kind of communities, the surplus production is almost nule, because the production/consumption ratio approaches zero, severely reducing the capacity of commercial exploitation.

4.1. The logistic curve approach

According to Graham (1939), the maximum yield that can be extracted from a wild stock is found at the half of the virgin size of that population, as seen in Fig. 1A, B. A similar view is commented by Zabel et al. (2003). After this premise, a simplistic approach can be adopted by assuming that when the catch trend shows a maximum, followed by a decline, then that

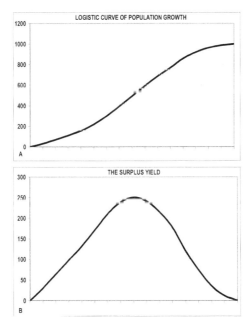

Figure 1. Principles of the logistic growth of a population (A) and the surplus yield of an exploited stock (B). Horizontal scale of Fig. A is time and in Fig. B indicates population size.

maximum yield corresponds to the half of the population size at the virgin stock. Stock assessment based upon this approach is very limited and despite that its ecological principles as background are valid, there are many factors constraining the validity of this procedure and therefore other approaches more accurate and based upon age structure have been adopted over time.

By following the former statements, a simple approach to roughly estimate the stock biomass is by just fitting a parabola to the catch records of some fisheries or regions, even deliberately ignoring a relationship of the stock density of populations, just by usually using the catch per unit of effort as an indicator of stock density. In this case, time was used as an indirect indicator of fishing effort, because the information on this variable is not easily available and because it is beyond the scope of this paper. Therefore, second degree regressions were used to several fisheries and regions just to have an idea on when the maximum yields, presumably equivalent to the Maximum Sustainable Yields (MSY), were attained. It is assumed that the stock biomass is at least twice bigger than the maximum yield attained in a certain time, and in that point is supposed that the exploitation rate E, is 50 per cent. This approach is conservative, because the intrinsic growth rate is not provided, given that many populations are involved. In the stock assessment process, the E value is usually lower than 0.5; however, by considering that many species are involved in the procedure is analysis, is likely to expect that in this collection there may be species which are overexploited, as well as others which may be underexploited. For this reason, it is reasonable to adopt a conservative criterion instead of being too optimistic assuming that the biomass could reach higher values. It is pertinent to mention that most of the regressions applied and described in the following paragraphs excepting three, provided high and significant R^2 coefficients.

On being consistent with this idea, estimations of the MSY by applying a parabola were fitted to catch data of the world fisheries exploited and recorded for different regions as shown in Fig. 2 (A - F) and in Table 1. The time scale of catch extracted from FAO (2010), data goes from 1950 to 2010. It is evident that in most cases the catch has attained a maximum yield, which for practical purposes; it can be considered as equivalent to the MSY level.

4.2. Biomass and fish production

The FAO (1995, 2005) is involved in the task of recording the world statistics of food production and often publishes assessments accounting for the status of world fisheries (FAO 1995, 2005; Froese and Pauly 2012). The catch records are grouped by statistical regions subdivided in 17 sub regions and in the following paragraphs, some highlights on the current status of the fisheries of these regions and sub regions is given, as well as some rough estimations of the biomass on which the exploitation of fish resources is based.

4.2.1. The Atlantic

In the Atlantic North-eastern, the MSY level was attained in the middle eighties (Fig. 2A), with 11.6 million (M) mt; this catch implies a biomass of 23.2 M mt with a significant

REGION	MSY	BIOMASS	Mean 2008 - 10 YIELD	Mean 2008 - 10 BIOMASS
ATLANTIC NORTHEASTERN	11,600,000	23,200,000	8,600,000	17,200,000
ATLANTIC EASTERN CENTRAL	3,700,000	7,400,000	3,750,000	7,500,000
ATLANTIC SOUTHEASTERN	2,700,000	5,400,000	1,300,000	2,600,000
ATLANTIC NORTHWESTERN	3,500,000	7,000,000	2,400,000	4,800,000
ATLANTIC SOUTHWESTERN	2,650,000	5,300,000	1,840,000	3,680,000
GULF OF MEXICO*	800,000	1,600,000	550,000	1,100,000
TOTAL ATLANTIC	**24,150,000**	**48,300,000**	**17,890,000**	**35,780,000**
PACIFIC NORTHEASTERN	2,750,000	5,500,000	2,910,000	5,880,000
PACIFIC NORTHWESTERN	22,550,000	45,100,000	20,900,000	41,800,000
PACIFIC WESTERN CENTRAL	12,000,000	24,000,000	12,000,000	24,000,000
PACIFIC EASTERN CENTRAL	2,000,000	4,000,000	2,000,000	4,000,000
PACIFIC SOUTHEASTERN	14,500,000	29,000,000	10,900,000	21,800,000
PACIFIC SOUTHWESTERN	800,000	1,600,000	600,000	1,200,000
TOTAL PACIFIC	**54,800,000**	**109,600,000**	**48,810,000**	**97,680,000**
ANTARCTIC INDIAN OCEAN	90,000	180,000	10,000	20,000
INDIAN OCEAN EASTERN	7,000,000	14,000,000	6,800,000	13,600,000
INDIAN OCEAN WESTERN	4,500,000	9,000,000	4,500,000	9,000,000
TOTAL INDIAN OCEAN	**11,590,000**	**23,180,000**	**11,310,000**	**22,620,000**
ANTARCTIC TOTAL	40,000	80,000	5,000	10,000
MEDITERRANEAN & BLACK SEA	1,700,000	3,400,000	1,500,000	3,000,000
OUTSIDE THE ANTARCTIC	80,000	160,000	20,000	40,000
TOTAL MARINE REGIONS	**99,710,000**	**199,420,000**	**79,565,000**	**159,130,000**

*Included in the Atlantic Southwestern region

Table 1. Maximum yields, equivalent to the MSY, of catch data recorded in FAO statistics for the seventeen statistical areas. Biomass estimates of total yields per area within a region and the total for the whole region are indicated. Current average yields, for the years 2008-2010 and their corresponding biomass are also shown on the two right side columns. Values are rounded, in mt.

decrease in biomass of 6 million mt in the last three years (Table 1). In Fig. 2B, the maximum catch of the Atlantic Eastern Central is displayed, and corresponds to 3.7 M mt, attained in the year 2000; this figure corresponds to a biomass of 7.4 M mt, but at the end of the period displays an increase of 100,000 mt. In the Atlantic South eastern, the maximum yield was obtained in the early eighties, with 2.7 M mt (Fig. 2C); the corresponding biomass is 5.4 M mt, with a significant decrease in biomass during the last three years to only 2.6 M mt. The catch trend of the Atlantic North western (Fig. 2D) is declining, with a maximum of 3.5 M mt attained in the early seventies; to this figure corresponds a biomass of 7 M mt (Table 1). The low biomass estimated for the years 2008-2010, with somewhat more than 4.8 M mt, is something to be concerned. The catch trend of the Atlantic South western (Fig. 2E) is not very clear, because it seems to attain a maximum followed by a decline, but the projection of the regression line suggests that the maximum yield will be reached until the year 2030 with

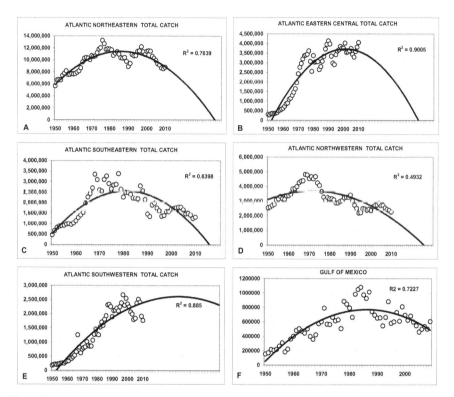

Figure 2. Trend of total catches extracted from several regions of the Atlantic in the period 1950 – 2010. A. Atlantic north eastern; in this region the maximum catches were obtained in the late eighties. B. Atlantic eastern central; the maximum yields were obtained around the year 2000. C. Atlantic south eastern; the maximum yield was obtained in the early eighties. D. Atlantic north western; the maximum yield was obtained by the year 1970, with a declining trend afterwards. E. Atlantic south western. It is not clear whether the maximum yield was attained by the early 2000's, or it still may grow to a maximum near the year 2030. In the Gulf of Mexico, whose data are included in those of Fig. 2.E, more than 60 species caught and recorded in the statistics, are included in this analysis; here, the MSY was attained in the middle 80's.

2.65 M mt. The corresponding biomass will be 5.3 M mt (Table 1); the stock current biomass is 3.68 M mt. It was possible to examine with some detail the catch trend of the Gulf of Mexico (Fig. 2F), whose values are part of those for the Atlantic South western; in this case, the maximum yield was obtained in the late eighties with 800,000 mt, with a corresponding biomass of 1.6 Million mt; the current biomass is only 1.1 M mt. The global MSY for the Atlantic Ocean is 24.15 M mt, corresponding to a biomass of 48.3 M mt but these values do not correspond to the same year; unfortunately in all cases but one, current yields were left behind and the current biomass is considerably lower than the figures provided. The current biomass estimated for the Atlantic Ocean amounts to 35.78 M mt (Table 1).

4.2.2. The Pacific

The catch obtained from this region at the maximum yield level, accounts to 62 per cent of world catch, with 54.85M mt (Table 1). The biomass from which this catch was extracted is 109.6 M mt. The current biomass is 89 per cent of the one at the MSY level. In Fig. 3A the catch trend of the Pacific north eastern is displayed; here the maximum yield was recorded by the year 2000, with almost 3 M mt, extracted from a stock biomass of 5.9 M mt; current biomass is unfortunately one Million lower and the trend is declining. In the Pacific north western, a similar trend is displayed (Fig. 3B), the maximum yield was obtained also by the year 2000, with nearly 22.6 M mt corresponding to a biomass of 45.1 M mt. The current

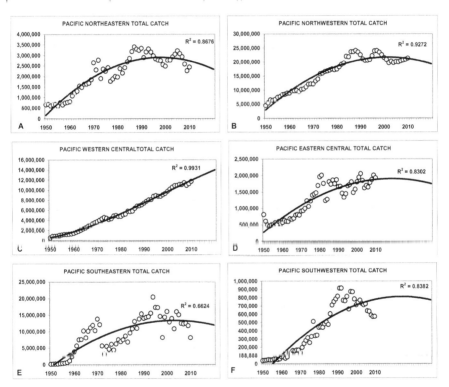

Figure 3. Trend of total catches extracted from several regions of the Pacific Ocean in the period 1950 – 2010. A. Pacific north eastern; in this region the maximum catches were obtained in the late nineties. B. Pacific north western; the maximum yields were obtained around the year 2000. C. Pacific western central; the maximum yield has not been reached and the fisheries seem to be in the eumetric phase. D. Pacific eastern central; the maximum yield seems that will be obtained in the near future. E. Pacific south eastern, the maximum yield was attained in the middle 2000's, but the catch of the last three years suggests a decline. F. Pacific south western, the trend suggests that the maximum yield has not been reached yet, but the catch has been declining since the last fifteen years.

biomass is 41.8 M mt. The catch of the Pacific western central displays a growing trend with 12 M mt in the last three years (Fig. 3C). To this catch corresponds a stock biomass of 24 M mt; no signs of stabilization of the catch are perceived, which is encouraging. In the Eastern central region, the yield seems to be attaining a maximum with around 2 M mt and a biomass of 4 M mt (Fig. 3D). These values are considered the current ones. The catch of the Pacific western central displays a growing trend with 12 M mt in the last three years (Fig. 3C). To this catch corresponds a stock biomass of 24 M mt; no signs of stabilization of the catch are perceived, which is encouraging. In the Eastern central region, the yield seems to be attaining a maximum with around 2 M mt and a biomass of 4 M mt (Fig. 3D). These values are considered the current ones. The south eastern region displays large variability, and the trend suggests that the maximum was attained a few years before, with a catch of 14.5 M mt corresponding to a biomass of 29 M mt. The mean catch of the last three years indicates a biomass decline to 21.8 M mt. The south western region suggests that the maximum was already attained, but the trend indicates that it will be reached within 15 years or so, with a catch of 800,000 mt and a biomass of 1.6 M mt; the current biomass is only 1.2 M mt.

4.2.3. The Indian Ocean, the Antarctic and the Mediterranean-Black Sea

These regions hardly attain a yield of 14 M mt at the MSY level, being the Indian Ocean the most productive of this group with 12 M mt caught in the whole area. The stock biomass approaches to 24 M mt at the MSY level but at the current exploitation level, this variable implies a reduction of almost 2 M mt, with almost 22 M mt.

The three regions in which it is divided show remarkable differences implying important characteristics in the fishing intensity applied; for instance, the Antarctic zone seems to have been completely overexploited and probably collapsed since 1992 (more recent catch data are not available) and the maximum catch was attained by the early eighties with nearly 100,000 mt as mean trend (Fig. 4A). The same as the Pacific western central, in the Indian ocean eastern the yield describes an increasing trend, with nearly 7 M mt in the last three years, with no signs of stabilization in the near future, which is also encouraging (Fig. 4B). The catch in the western region is also growing, but it seems to be stabilizing currently; the catch is 4.5 M mt corresponding to a maximum stock biomass of 9 M mt (Fig. 4C).

The catch at the Antarctic shows a declining trend, with a maximum yield of nearly 40,000 mt recorded in the middle fifties (Fig. 4D) and a stock biomass of 80,000 mt. Same as the Antarctic Indian Ocean, these fisheries seem to be completely collapsed since the late eighties.

The Mediterranean and Black sea display a quite stable catch trend through the last 30 years and the MSY was attained by the year 1990 with 1.6 M mt, from a biomass of 3.2 M mt. After that year there has been a slow declining trend, such that the current stock biomass is no higher than 3 M mt (Fig. 4E).

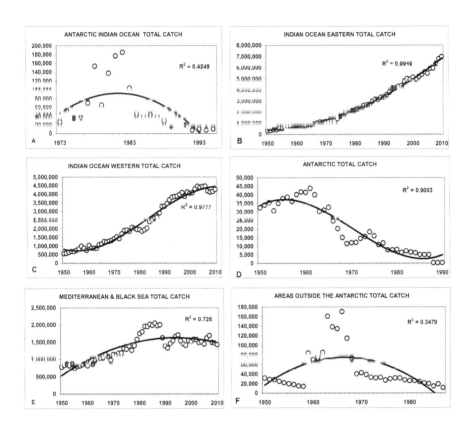

Figure 4. Trend of total catches extracted from several regions of the Indian Ocean, Antarctic, and Mediterranean-Black Sea in the period 1950 - 2010. A. Antarctic Indian ocean; the maximum yield was obtained in the year 1980 and the fisheries collapsed by the early nineties. B. Indian Ocean eastern; the fisheries are in a eumetric stage of growth nowadays. C. Indian Ocean western; after a sustained growth since the fifties, these fisheries appear to be approaching their MSY level nowadays. D. Antarctic ocean; all information related to this ocean confirms a collapse of its fisheries, as occurred in this case. E. Mediterranean-Black sea; after a period of slow but consistent growth, the MSY appears to have attained the MSY level in the late nineties, followed but a slow decline nowadays. F. The areas outside the Antarctic show the same trend as the Antarctic itself, with an apparent collapse nowadays.

Finally, the areas outside the Antarctic, apart from a peak of the catch in the middle sixties, display low yields that currently are above 11,000 mt from a stock biomass of 40,000 mt. This fishing region appears to be collapsed too.

5. Concluding remarks

The use of surplus yield models for assessment of exploited fish stocks, has becoming an tool hardly used nowadays, because the use of age structured methods with the aid of computing techniques, allow more powerful and more accuracy in the assessments. There were times when fisheries researchers devoted their efforts into that approach and more sophisticated variations of the original statements were made (Walter 1975, 1978; Csirke and Caddy 1983; Arreguín-Sánchez and Chávez 1986; Polacheck et al. 1993; Fréon and Yáñez 1995); however, this approach has became obsolete over time, despite its background ecological principles are still valid. However, the large variance implicit in the estimations caused by several factors, contributed in a great deal to its current lack of use. Despite this consideration, it was decided to adopt that approach in this paper, for several reasons, the first one is it accessibility and easy way to just fitting a second degree curve in the spreadsheet where a bunch of catch data involving as many species as they are exploited in the world oceans, just to have a guideline on the maximum yield level and the year when it was reached. It also provided a minimum basic requirement for the estimation the stock biomass on which fisheries of each region were based.

It is remarkable to realize that the maximum yield of the world oceans approaches very close to 100 M mt and the biomass of all the exploited stocks is near to 200 M mt. Another important point to call the attention is that in most cases, the MSY was attained more than a decade ago and that the current yield and stock biomass are nearly 40 per cent below those maxima. This is something to concern and is a possible indicator of excessive pressure on the fish stocks and in this respect those on the Antarctic seem to be the most heavily impacted by fishing activities. Evidently the over exploited fisheries have passed by several stages (Pauly et al. 1998) already pointed by other authors (Harding 1968; Feeny et al. 1990; Myers and Worm 2003) and unfortunately the perspective suggests that other world oceans apart from the Antarctic, will follow the same steps if no action is taken by the nations to ensure exploiting the sea in a sustainable way (Jorgensen et al. 2007).

Author details

Ernesto A. Chávez[*] and Alejandra Chávez-Hidalgo
Centro Interdisciplinario de Ciencias Marinas, Instituto Politécnico Nacional, La Paz, México

6. References

Arreguín-Sánchez F. and Chávez E.A. 1986. Influencia del reclutamiento sobre el rendimiento pesquero. IOC Workshop Report No. 44:95-104

Botsford L.W., Castilla J.C., and Peterson C.H. 2012. The management of fisheries and marine ecosystems. Science 277:509-515

---

[*] Corresponding Author

Csirke J. and Caddy J.F. 1983. Production modeling using mortality estimates. Can. J. Fish. Aquat. Sci. 40: 43-51.

Christensen, V. and Pauly D. 1993. Trophic models of aquatic ecosystems. International Center for Living Aquatic Resources Management. Manila, Philippinnes.

Cushing D.H. and Walsh J.J. 1976. The ecology of the seas. Blackwell.

FAO. 1995. http://www.fao.org/docrep/006/ac442s/AC442s31.htm# (Declaration Kyoto Conference/Fisheries FAO Fisheries Department).

FAO. 2005. Review of the state of world marine fishery resources. Marine resource service Fishery resource division. FAO Fisheries Department. Technical paper 457.

Feeny D., Berkes F., MacCay B.J., and Acheson J.M. 1990. The tragedy of the commons: twenty-two years later. Human Ecology, 1(18):1-19.

Fréon P. and Yáñez E. 1995. Influencia del medio ambiente en evaluación de stock: una aproximación con modelos globales de producción. Invest. Mar. Valparaíso, 23:25-47.

Friedland K.D., Stock C., Drinkwater K.F., Link, J.S., Leaf R.T., Shank B.V., Rose J.M., Pilskain C.H., and Fogarty M.J. 2012. Pathways between primary production and fisheries yields of large marine ecosystems. PloS One 7(1):e28945

Froese, R. and D. Pauly Editors 2012. FishBase. World Wide Web electronic publication. http://www.fishbase.org, version (08/2011).

Graham M. 1939. The sigmoid curve and the overfishing problem. Rapp. Conseil. Explor. Mer. 110:15-20.

Harding G. 1968. The tragedy of the Commons: The population problem has no technical solution; it requires a fundamental extension in mortality. Science 162: 1243-1248.

Jorgensen C., Enberg K., Dunlop E.S., Arlinghaus R., Boukal D.S., Brander K., Ernande B., Gårdmark A., Johnston F., Matsumura S., Pardoe H., Raab K., Silvia A., Vainikka A., Dieckmann U., Heino M., and Rijnsdrop A.D. 2007. Managing evolving fish stocks. Science 318:1247.

Margalef R. 1974. Ecología. ed. Omega, 951 pp.

Myers R.A. and Worm B. 2003. Rapid worldwide depletion of predatory fish communities. Nature 423:280-283.

Pauly D. and Christensen, V. 1995. Primary production required to sustain global fisheries. Nature 376:255-279.

Pauly D. Christensen V., Dalagaard J., Froese R., and Torres F.Jr. 1998. Fishing down marine food webs. Science 279:860-863.

Polacheck T., Hilborn R., and Punt A.E. 1993. Fitting surplus production models: comparing methods and measuring uncertainty. Can. J. Fish. Aquat. Sci. 50: 2597-2607.

Russell-Hunter W.D. 1970. Aquatic productivity. Macmillan Pub. Co., NY., 306 pp.

Walter G.G. 1975. Graphical method for estimating parameters in simple models of fisheries. J. Fish. Res. Board Can. 32: 2163-2168.

Walter G.G. 1978. A surplus yield model incorporating recruitment and applied to a stock of Atlantic mackerel (Scomber scombrus). J. Fish. Res. Board Can. 35:229-234.

Zabel R.W, Harvey C.J, Katz S.L, Good T.P., and Levin P.S. 2003. Ecologically sustainable yield. American Scientist. 91(2):150-157.

Chapter 6

Microbial Biomass in Batch and Continuous System

Onofre Monge Amaya, María Teresa Certucha Barragán and Francisco Javier Almendariz Tapia

Additional information is available at the end of the chapter

1. Introduction

Microorganism is a microscopic organism, commonly used term to describe a cell or more, including viruses [1]. This definition includes all prokaryotes, as the eukaryotic unicellular: protozoa, algae and fungi. Bacteria are an extremely diverse group of organisms with extensive variation of morphological, ecological and physiological, which due to their diversity, are regularly found in heterogeneous communities [2].

The microorganisms are widely used in the biological treatment of waste solids and liquids [3]. Importantly, the microorganisms used in biological treatments as a partnership a community or consortium [2].

Nowadays there is a current increasingly important in the sense of using microorganisms, especially bacteria, algae and fungi, for decontamination and to help recovery from natural environments and for treating municipal or industrial effluents. It is estimated that the best microorganisms for the removal of toxins present in a place, are initially isolated in their own area where they have been naturally selected, although in a second genetic manipulation of these can significantly strengthen the capacity of microorganisms culture collections and isolates from the different environments of interest. This is supported by the observation that microorganisms capable of living in a polluted environment, and thus perform vital functions, in cellular metabolism have highly effective devices for decontamination. To know the details of these mechanisms could be exploited to purify the water, such as using bacteria capable of capturing heavy metals on their cell wall [4].

Thus, this rapid advance of science and technological development has allowed decontamination using microorganisms in the water. Using the metabolism of microorganisms has enabled the construction of biological reactors at much lower cost than physicochemical, and also the construction of treatment plants mixed system for greater efficiency [5]. Developed countries currently use these biomining processes through the involvement of bacteria such as *Acidithiobacillus ferrooxidans*.

Some species of *Klebsiella* and *Pseudomonas* are capable of degradation of reactive pollutants. It also recognizes the ability of some microorganisms or their enzymes to degrade under certain conditions to cyanide reagent, employed in the leaching of gold and silver recovery [6]. Several studies have shown that the biomass of different species of bacteria, fungi and algae are capable of concentrating metal ions in their structures that are found in aquatic environments [7].

Bioremediation is defined as a natural process, during in which different microorganisms are capable of removing organic and inorganic contaminants in a given environment [8]. In the bioremediation process, there are different objectives to assess; mainly seeks to avoid a long term noxious effects to other organisms as well as natural resources, it seeks to recover the ecological balance that exists in the environment; and finally, it seeks to achieve that the contaminated area, with a subsequent treatment by biological processes, can be reused for recreation or productive purposes [9].

Different types of biological treatment systems are used in the field of environmental engineering. The biochemical reactions leading to the oxidation of organic matter are conducted in reactors that can be classified as aerobic or anaerobic, suspended growth or biofilm, with mechanical or without mechanical mixing, etc. In order to design an appropriate reactor for a given wastewater treatment system, both the microbial kinetics of substrate removal and the fundamental properties of different reactors have to be understood [10].

The use of microorganisms as tools of decontamination is fairly recent. The biggest advances in the field were made after the oil spill of the Exxon Valdez on the coast of Prince William, in Alaska (1990). Since this ecological disaster lot of oil left in the water, sought alternative ways of dealing with pollution.

The scientists who developed the first successful experiences of bioremediation of oil a large scale in Alaska, were based on the premise that all natural ecosystems have organisms capable of metabolizing toxic compounds and xenobiotics, although these are often found in proportions less than 1% microbial community. This premise was fulfilled in Alaska and in almost all cases studied later [5].

Several studies have shown that the biomass of different species of bacteria, fungi and algae are capable of concentrating metal ions in their structures that are found in aquatic environments [7]. Bacteria as the genus *Pseudomonas* of mining environments have been identified that are resistant to heavy metals such as cadmium (Cd), copper (Cu) and lead (Pb) [11]. Some species of marine microalgae, *Staphylococcus saprophyticus* and fungi have been reported the biosorption of cadmium, chromium, lead and copper from wastewater [7, 12, 13].

In other studies it is known that microbial strains are able to bioremediate contaminated soils with different metals and organic compounds. It is known that *Escherichia coli* is able to bioaccumulate cadmium concentrations of 5 mg/L as well as copper and zinc which are taken from the culture medium by a process in which occurs a binding peptides secreted by the bacteria [14]. Other studies reported bacterial consortia isolated from mining effluents that adsorbed copper [15], as well as anaerobic consortia that in continuous system showed high percentages of copper and iron removal [16, 17].

It is well known that the use of the water drainage basins are the primary sources for anthropogenic unloads; it represents a risk for the human health, particularly the pollution caused by the high concentrations of some heavy metals, such as, zinc (Zn), nickel (Ni), chromium (Cr), lead (Pb) and copper (Cu) [18, 19, 20]. Those metals go through the aquatic environment principally by direct loads of industrial sources, soils and sediments and are distributed in the water, biota, being the mining industry one of the most important source [21, 22, 23]. An example of contamination by heavy metals in water and sediment, it is the San Pedro River basin, one of the most important rivers of the north of Sonora, Mexico. This river has been severely contaminated by the wastewater discharges with high concentration of copper generated during the mining activity of the region. In addition, wastewaters discharges untreated raw sewage coming from the city of Cananea, Sonora [24]; also contribute with the pollution of this source of water. Considering the importance and impact of this problem on the community of Cananea city, different kinds of researches had been developing. In reports the sampling stations in the San Pedro River and the propagation and isolation of the bacteria and the copper biosorption in an aerobic bioreactor. However, it is necessary to make more studies that allow suggesting other alternatives for the treatment of acid mine drainages (AMD), and consequently, reducing the concentration of the heavy metals in the San Pedro River, until acceptable levels according to Mexican regulations (NOM-001])[15).

2. Anaerobic and aerobic process

2.1. Anaerobic processes

Opposite to the aerobic are anaerobic processes, which are performed in the absence of oxygen by groups of heterotrophic bacteria, which in a process of liquefaction/gasification in two stages, becomes a 90% organic matter present at first in intermediate (partially finished products stabilized that include organic acids and alcohols) and then to methane and gaseous carbon dioxide:

$$organic\ mater \xrightarrow{acid\ forming\ bacteria} intermediaries + CO_2 + H_2S + H_2O \qquad (1)$$

$$organic\ acids \xrightarrow{methane\ forming\ bacteria} CH_4 + CO_2 \qquad (2)$$

The process is applied universally in hot anaerobic digesters, where in the primary and biological sludge is maintained for about 30 days at 35 °C to reduce its volume (about 30%)

and their ability to putrefaction, there by simplifies the removal of sludge. The advantage of this type of digestion is that generates energy in the form of methane and the production of sludge is only 10% [5, 25].

2.2. Anaerobic treatment

Anaerobic treatment processes require the presence of a diverse closely dependent group of bacteria to bring about the complete conversion of complex mixtures of substrates to methane gas. It is puzzling that single species of bacteria have not evolved to convert at least simple substrates such as carbohydrates, amino acids, or fatty acids all the way to methane [26].

Conventional phase and high-rate two-phase anaerobic digestion processes have frequently been employed in order to treat both soluble and solid types of domestic and industrial wastes. The most significant outcome of anaerobic digestion processes is that they generate energy in the form of biogas namely, methane and hydrogen. Therefore, due to current imperative environmental issues such as global warming, ozone depletion, and formation of acid rain, substitution of renewable energy sources produced from biomass, such as methane and hydrogen, produced through anaerobic digestion processes will definitely affect the demand and consumption of fossil-fuel derived energy [27].

The anaerobic treatment is a biological process widely used in treating wastewater. When these have a high organic load, presents itself as the only alternative would be an expensive aerobic treatment, due to the oxygen supply. The anaerobic treatment is characterized by the production of "biogas" consisting mainly of methane (60-80%) and carbon dioxide (40-20%) and capable of being used as fuel for generating thermal energy and / or electric. Furthermore, only a small fraction of COD treated (5-10%) is used to form new bacteria, compared to 50-70% of an aerobic process. However, the slow anaerobic process requires working with high residence times, so it is necessary to design reactors or digesters with a high concentration of microorganisms [28, 29]. Actually is a complex process involving several groups of bacteria, both strictly anaerobic and facultative, which, through a series of stages and in the absence of oxygen, flows mainly in the formation of methane and carbon dioxide. Each stage of the process, described below, is carried out by different groups of bacteria, which must be in perfect balanced. Figure 1 shows a schematic representation of the main conversion processes in anaerobic digestion, suggested by Gujer [30].

(1) Hydrolysis. In this process complex particulate matter is converted into dissolved compounds with a lower molecular weight. The process requires the mediation of exo-enzymes that are excreted by fermentative bacteria. Proteins are degraded via (poly) peptides to amino acids, carbohydrates are transformed into soluble sugars (mono and disaccharides) and lipids are converted to long chain fatty acids and glycerine. In practice, the hydrolysis rate can be limiting for the overall rate of anaerobic digestion. In particular the conversion rate of lipids becomes very low below 18°C.

(2) Acidogenesis. Dissolved compounds, generated in the hydrolysing step, are taken up in the cells of fermentative bacteria and after acidogenesis excreted as simple organic compounds like volatile fatty acids (VFA), alcohols and mineral compounds like CO_2, H_2, NH_3, H_2S, etc. Acidogenic fermentation is carried out by a diverse group of bacteria, most of which are obligate anaerobe. However, some are facultative and can also metabolize organic matter via the oxidative pathway. This is important in anaerobic wastewater treatment, since dissolved oxygen (DO) otherwise might become toxic for obligate anaerobic organisms, such as methanogens [31].

(3) Acetogenesis. The products of acidogenesis are converted into the final precursors for methane generation: acetate, hydrogen and carbon dioxide. As indicated in Figure 1, a fraction of approximately 70% of the COD originally present in the influent is converted into acetic acid and the remainder of the electron donor capacity is concentrated in the formed hydrogen. Naturally the generation of highly reduced material like hydrogen must be accompanied by production of oxidized material like CO_2.

(4) Methanogenesis. Methanogenesis may be the rate liminting step in the overall digestion process, especially at high temperatures ($> 18°C$) and when the organic material in the influent is mainly soluble and little hydrolysis is required. Methane is produced from acetate or from the reduction of carbon dioxide by hydrogen using acetotrophic and hydrogenotrophic bacteria, respectively:

$$Acetotrophic\ methanogenesis: CH_3COOH \rightarrow CH_4 + CO_2 \qquad (3)$$

$$Hydrogenotrophic\ methanogenesis: 4H_2 + CO_2 \rightarrow CH_4 + 2H_2O \qquad (4)$$

Different from aerobic treatment where the bacterial mass was modeled as a single bacterial suspension, anaerobic treatment of complex wastewater, with particulate matter in the influent, is only feasible by the action of a consortium of the four mentioned groups of bacteria that each have their own kinetics and yield coefficients. The bacteria that produce methane from hydrogen and carbon dioxide grow faster than those utilizing acetate that the acetotrophic methanogens usually are rate limiting for the transformation of acidified wastewaters to biogas [32].

The different groups of bacteria involved in the conversion of influent organic matter all exert anabolic and catabolic activity. Hence, parallel to the release of the different fermentation products, new biomass is formed associated with the four conversion processes described above. For convenience, the first three processes often are lumped together and denominated acid fermentation, while the fourth step is referred to as methanogenic fermentation.

The removal of organic matter-COD during the acid fermentation is limited to the release of hydrogen only 30% of the organic matter is converted into methane via the hydrogenotrophic pathway. Hence, a necessary condition for efficient organic matter removal in an anaerobic treatment system is that a sufficient mass of acetotrophic methanogens develops.

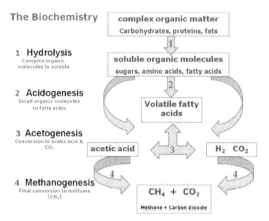

Figure 1. Schematic representation of the main conversion processes in anaerobic digestion

Acid fermentation tends to cause a decrease in the pH because of the production of VFA and other intermediates that dissociate and produce protons. As methanogenesis will only develop well at a neutral pH values, instability may arise, if for some reason the rate of acid removal by methane production falls behind the acid production rate: the net production of acid will tend to cause a decrease in pH, and thus may reduce the methanogenic activity further. In practice, this so called "souring" of the anaerobic reactor contents is the most common cause for operational failure of anaerobic treatment systems. The danger of souring can be avoided, by maintaining the proper balance between acid and methanogenic fermentation which in fact means that both the methanogenic digestion capacity and buffer capacity of the system should be sufficiently high [29, 33].

2.3. Chemical oxygen demand (COD)

Natural organic detritus and organic waste from wastewater treatment plants, failing septic systems, and agricultural and urban runoff, acts as a food source for water borne bacteria. Bacteria decompose these organic materials using dissolved oxygen. The determination of Chemical Oxygen Demand (COD) is widely used in municipal and industrial laboratories to measure the overall level of organic contamination in wastewater. The contamination level is determined by measuring the equivalent amount of oxygen required to oxidize organic matter in the sample.

In the COD method, the water sample is oxidized by digesting in a sealed reaction tube with sulphuric acid and potassium dichromate in the presence of a silver sulphate catalyst. The amount of dichromate reduced is proportional to the COD. A reagent blank is prepared for each batch of tubes in order to compensate for the oxygen demand of the reagent itself.

Over the range of the test a series of colors from yellow through green to blue are produced. The color is indicative of the chemical oxygen demand and is measured using a photometer. The results are expressed as milligrams of oxygen consumed per liter of sample [34].

Test Procedure

i.   Mixed reagent. Potassium dichromate solution. To a 1000 mL volumetric flask, add 42.256 ± 0.001 g of potassium dichromate (previously dried for one hour at 140 - 150 °C). To the flask, add approximately 500 mL of water and mix the contents to dissolve. Then add 33.3 g de $HgSO_4$ to the potassium solution. Add in an ice bath slowly 167 mL pure H_2SO_4. When the mixture has cooled, stir the mixture until the solid dissolves and dilute to one liter.

ii. Silver sulphate in sulphuric acid (10 g). To a glass bottle, add 10.0 ± 0.1 g of silver sulphate and 1000 ± 10 mL of sulphuric acid and stopper. To obtain a satisfactory solution, swirl the initial mixture and allow it to stand overnight. Swirl the contents again until all the silver sulphate dissolves. This solution may be stored in the dark at room temperature for up to an indefinite period.

iii. Pipette 2.0 mL sample into cuvette with 2.0 mL of the mixed reagent solution and 1.0 mL of the silver sulphate in sulphuric acid solution. Invert cuvettes carefully.

iv. Heating reactor for determination of COD for about 30 minutes.

v.   Heat cuvettes for 2 h at 150 °C.

vi. read cuvettes in the spectrophotometer at 620 nm.

2.4. Technique to determine alkalinity

Typical control strategy in methanogenic anaerobic reactors is to maintain a relatively low concentration of volatile fatty acids (VFA) and a pH range of 6.6 < pH < 7.4. Normally in such reactors the carbonate system forms the main weak-acid system responsible for maintaining the pH around neutrality, while the VFA systems (acetic, propionic, and butyric acids) are the major cause for pH decline. Under stable operating conditions, the H_2 and acetic acid formed by acidogenic and acetogenic bacterial activity are utilized immediately by the methanogens and converted to methane. Consequently, the VFA concentration is typically very low, carbonate alkalinity is not consumed and the pH is stable. Conversely, under overload conditions or in the presence of toxins or inhibitory substances, the activity of the methanogenic and acetogenic populations is reduced causing an accumulation of VFA which in turn increases the total acidity in the water, reducing pH. The extent of the pH drop depends on the H_2CO_3 alkalinity concentration. In medium and well-buffered waters (typically the case in anaerobic digestion), high concentrations of VFA would have to form in order to cause a detectable pH drop, by which time reactor failure would have occurred. Therefore, pH measurement cannot form the sole control means, and direct measurement of either (or both) VFA or H_2CO_3 alkalinity concentration is necessary.

The most used technique for the determination of alkalinity for the control of the system anaerobic is described below:

25 mL of sample are taken and placed on a plate with stirring to a solution titrated with 0.02 N sulfuric acid, it initial pH is measured and the acid is added until the pH changes to 5.75 volume of spent acid, followed by titrating until the pH changes to 4.3 and the volume of spent acid is taken and is determined the alpha value.

Alpha = acid vol. (5.75)/ acid vol. (4.3) if this value is greater than 0.55 the bioreactor is acidified and must add a buffer, on the contrary, if it is less, acid must be added [35].

2.5. Methanogenic activity determination

The specific methanogenic activity (SMA = $gDQO\text{-}CH_4 \cdot gVSS^{-1} \cdot d^{-1}$) is defined as the rate of methane production, expressed as COD, regarding biomass expressed as the content of volatile suspended solids (VSS). In anaerobic degradability test measures the rate of degradation of a compound relative to a standard compound that is acetic acid determining [36].

$$SMA = \frac{m}{\gamma_{CH_4} \cdot X}$$ (5)

Where: Slope = $m = \frac{LCH_4}{d}$; Biomass = X [=] gVSS/L; Methane conversion $\gamma \frac{CH_4}{COD} = 0.35 \frac{LCH_4}{gCOD}$

Methanogenic activity and toxicity

Methanogenic activity were performed using the pressure transducer technique, which involves the monitoring of the pressure increase developed insealed vials fed with non-gaseous substrates or pressure decrease in vials pressurised with gaseous substrates. Strict anaerobic conditions must be maintained. The same technique can be used to perform the methanogenic toxicity tests. The fifty percent inhibition concentration (IC_{50}) was defined as the methanogenic concentration that caused a 50% relative activity loss [35, 37].

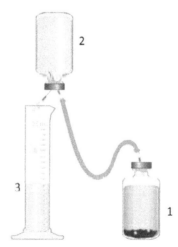

Figure 2. Schematic representation of methane measuring displacement of a solution of 3% NaOH. 1) serological bottle, 2) serological bottle with 3%NaOH 3) test tube to measure the displaced NaOH

The technique is as follows:

1. The sludge is left in mineral medium for 24 hours at 30-35 °C in order to consume the entire carbon source which may have been brought into water of the plant.
2. Methanogenic activity tests were conducted in 160 mL in serology bottles with an operating volume of 150 mL. The volume of volatile suspended solids was set at 2 g/L and COD concentration used was varied from 0.25, 0.5, 1, 2, 3 y 5 g/L using acetate as a carbon source, and staying a relationship of 0.125, 0.25, 0.5, 1, 1.5 y 2.5 gCOD/gVSS respectively.
3. The bottles were sealed with rubber stoppers, and incubated 24 h at 35°C.
4. Methane was determined by the displacement volume of a solution of 3% NaOH [38].

Figure 2 show a schematic representation of methane measuring.

2.6. Characterization of the sludge (total suspended solids (TSS), volatile suspend solids (VSS), fixed solids (FS))

The total solids (TS) contents of sludge are used in the design and process control of wastewater treatment facilities. Total dissolved solids (TDS) are used to evaluate the suitability of water for both domestic supplies and industrial purposes. The total suspended solids (TSS), including the volatile fraction (VSS), are commonly monitored to evaluate the degree of pollution in natural waters and serves as a key process control parameter for wastewater treatment operation.

The measurement of solids is by means of the gravimetric procedure. The various forms of solids are determined by weighing after the appropriate handling procedures.

1. Placing a clean porcelain crucible for 60 min at 550 °C in the muffle then is passed to a desiccator to cool and then weighed (noted as weight of the crucible)
2. Add with a blunt pipette 10 mL of anaerobics sludge. In a warming rack remove all the water possible. Go to the muffle furnace at 100 °C within 2 h; after this time is cooled and weighed (weight recorded as 100 °C).
3. Place the crucible in the muffle at 550 °C for 1 h and pass it to the desiccator and record the weight of the crucible (as weight at 550 °C).
4. Finally returned to the desiccator. After obtaining the three weights, and proceeded to determine the total solids, volatile and fixed as follows. [35]

$$Total\ suspended\ solids \qquad TSS: \frac{Weight\ to\ 100^{\circ}C - Weight\ of\ the\ crucible}{Sample\ volume}[=]\frac{g}{L} \qquad (6)$$

$$Fixed\ solids \qquad FS: \frac{Weight\ to\ 550^{\circ}C - Weight\ of\ the\ crucible}{Sample\ volume}[=]\frac{g}{L} \qquad (7)$$

$$Volatile\ suspended\ solids\ VSS: TSS - FSS[=]\frac{g}{L} \qquad (8)$$

2.7. Setting sludge volume index (SVI)

The sludge volume index is defined as 'the volume in mL occupied by 1 g of sludge after it has settled for a specified period of time' generally ranging from 20 min to 1 or 2 hr in a 1 – or 2 L cylinder. One-half hour is most common setting time allow the mixed liquor to settle for 30 min. (larger cylinder is desirable to minimize bridging of sludge floe and war effects). SVI is 50-150 mL/mg, the sludge settle ability if good.

SVI typically is used to monitor settling characteristics of activated sludge and other biological suspension. Although SVI is not supported theoretically, experience has shown it to be useful in routine process control. The SIV determination consisted of:

1. Place in the imhoff cone of 1000 mL, 100 mL of sludge and diluted to 1000 mL with phosphate buffer.
2. Was allowed to stand for 45 minutes and then stir the contents with a glass rod.
3. The volume occupied by the mud was measured by sedimentation after 30 minutes.
4. The SIV was calculated by dividing this volume by the present VSS g in 100 mL of sludge (Sludge/gVSS) [34].

2.8. Granule density

Among all the different types of anaerobic digesters applied at full scale, UASB (Upflow Anaerobic Sludge Blanket) reactors present the best commercial acceptance. The success of these reactors is related to their capacity for biomass accumulation by settling without the need of a carrier. Good settling properties are obtained through the flocculation of the biomass in the form of dense granules with diameters up to several millimetres. Actually, as individual cells and granules have similar densities, the greater settling velocity of the latter is only related to its larger particle size. The study of this phenomenon has lead to the development of several techniques for characterizing the resistance of the granules, their porosity, settling properties, bacterial composition and organization, activity, nature and composition of exopolymers, as well as their size distribution. This last parameter is particularly useful for studying the physico-chemical factors promoting sludge granulation [38].

1. Search about 6 stainless steel screens with an aperture of about 2 to 0.149 mm for the test.
2. In a vertical stack such that always at the top this larger diameter with respect to the bottom.
3. Take and pass a sample of 25 mL of sludge by the sieves.
4. Washing the sludge with a buffer and phosphate to make them pass through the screens; separated by size.
5. Retrieve the granules, separately, to be retained in the meshes with a backwash of phosphate buffer solution (Table 1) and then determine VSS, FS and TSS [38].

Compound	g/L
K_2HPO_4	4
$Na_2HPO_4 \cdot 7H_2O$	5.09
KH_2PO_4	1.08
pH solution 7.5	

Table 1. Phosphate buffer solution

3. Aerobic processes

Among the biological processes used for the construction of bioreactors, there are two fundamental types of processes: the aerobic and anaerobic.

Aerobic processes are those that need oxygen. There are strict aerobic processes, which are those that can only work if there is oxygen, and facultative aerobic processes, which are those that can switch to anaerobic, according to the concentration of oxygen available.

In general, the aerobic processes have the following reaction:

$$organic\ materia + O_2 \rightarrow CO_2 + H_2O + new\ cells \qquad (9)$$

As can be seen in the above reaction, essentially, aerobic metabolism is responsible for catalyzing larger molecules into carbon dioxide, water and new cells. It is noteworthy that the different groups of microorganisms have different metabolisms, and therefore are able to catalyze a wide range of substances, although sometimes other secondary products are obtained in addition to those mentioned.

Aerobic processes are very efficient, operate at a wide range of possible substances to degrade, and in relatively simple cycles are stable; there is rapid conversion of organic pollutants in microbial cells and their operation relatively free of odors [38].

3.1. Growth of microorganisms

3.1.1. Isolation

Most of classical and clinical microbiology depends on the isolation of a pure culture that consists of only one species. This isolate is then later used for characterization such as species determination or an antibiotic resistance profile. For many applications it is essential to isolate and maintain a pure culture of the organism of interest. The goal is to obtain isolated colonies of the organism of interest. These colonies arise from one single cell and are therefore a clone of that original cell. The original cell is called a colony-forming unit (CFU). To obtain a pure culture, it is crucial to maintain a sterile environmental.

To accomplish the microbiological analysis and isolation of strains, water samples are collected, this should be in sterile plastic containers of 500 mL. Is performed enrichment in nutrient broth, to ensure a favorable conditioning bacteria that may be stressed by environmental conditions and ensure better isolation, for which they are placed 10 mL of sterile nutrient broth sample, duplicate it.

Incubate at 37 ° C for 24 hours and reseed poured plate technique or grooves in the selected specific culture media: nutrient agar for *Bacillus sp* and *Pseudomonas sp*; agar EMB (eosin methylene blue) for *Enterobacteriaceae*; agar M17 for *Enterococcus*. Also plantings from water samples directly on agar PDA (Potato Dextrose Agar) acidifying the medium with tartaric acid for the isolation of fungi and yeast, and incubated at 22 °C [38].

3.2. Macroscopic and microscopic characterization

All cells contain the same mayor macromolecules in approximately the same proportions which perform essentially to same functions. Cell shape and arrangement are the initial steps for identifying bacteria. Cell grouping is related to the number of division and whether the cells remain together or separate after division. The Gram stain is an important toll for the classification and eventual identification of bacterial species.

The macroscopic characterization of the colonies was determined according to the observation of general appearance: shape, color, size, texture, elevation and range. The microscopic characterization to observe shape, color and size was performed by Gram stain for bacteria, with lactophenol blue stain for fungi and solution saline for yeast [37, 39].

3.3. Characterization of strains

3.3.1. Biochemical characterization and selection of strains

To grow, microorganisms from the environment should take all the substances required for the synthesis of their cellular materials and power generation. These substances are known nutrients. A culture medium should contain, therefore, all the necessary nutrients in appropriate amounts in the specific requirements of the microorganisms to what has been devised. At selected strains were performed the following biochemical tests, in order to know their identification: catalase production, nitrate reduction, mobility, indole production, the use of citrate as a carbon source, production of urease, methyl red, Voges Proskauer, carbohydrate fermentation, starch hydrolysis, gelatin hydrolysis and hydrolysis of esculin [40,41]. To determine resistance to low pH, this is changed in the culture media from 3 to 6. The pH was measured with a potentiometer, is adjusted with 10 M NaOH (sodium hydroxide) and HCl (hydrochloric acid).

3.4. Growth kinetics

The relatively large surface area of microorganisms exposed to the environment where they live enables them to take up and assimilate nutrients readily and a to multiply at impressive rates. An important parameter is the specific growth rate, wich is often expressed as the mean doubling time, defined as the time required by the microbial population to douyble its cellular protein content. The doubling time varies widely depending on the microbial species, the nature of the substrate and the degree of adaption to the substrate.

The growth of a bacterial culture can be determined by measuring the increase of turbidity in the medium as optical density or OD. The most common way to do this is to compare the absorbance of the culture to inoculated medium by shining light with a wavelength of 600 nm through the culture. The more growth occurs, the more turbid the culture will become and more light will be absorbed. Using optical density gives indirect measurement of bacterial growth. It doesn't tell anything about how many living cells are in the culture. This becomes especially important in stationary phase. Dead cells still absorb light. To determine the actual number of live bacteria the broth is diluted and plated on appropriate media and incubated. In theory, colonies arising on a plate originate from one single bacterium and give therefore an accurate number of the live cells in the culture at that time point.

Procedure:

The kinetics of selected bacterial growth, it is used 10 mL of 24 h culture of the strains and inoculated into 60 mL of nutrient broth, the following conditions: 35 °C and 100 rpm agitation. Samples were read every 30 minutes in a Spectronic 20D⁺ visible spectrophotometer at 600 nm, substituting readings transmittance (% T) in the following equation:

$$A = 2 - \log_{10}(\%T) \tag{10}$$

Where A, is the absorbance.

The different phases of growth kinetics can be observed by plotting the log (% T) versus time [1].

3.5. Biosorption process with bacteria in batch system

Microbial growth and substrate utilization expressions can be incorporated into mass balances to yield equations that can be used to predict effluent microorganism and substrate concentrations, and thus process efficiency. Continuous flow systems are grouped into two broad categories, suspended-growth and attached-growth processes, depending on whether the process microorganisms are maintained in suspension, or are attached to an inert medium (e.g., rocks, sand, granular activated carbon, or plastic materials). Attached-growth processes are also called fixed-film processes or biofilm processes.

Biosorption has provided an alternative process to the traditional physico-chemical methods, utilizing inexpensive biomass to sequester toxic heavy metals. In the 80 last decades, many researchers have focus on the treatment of wastewater containing heavy metals by the use of living organisms and/or their biomass. Many types of organisms such as bacteria, fungi, yeast and algae or their biomasses, have been used for metal uptake [42].

Biosorption tests batch system are carried out with each strain selected in 500 mL erlenmeyer flask, add 90 mL of a solution containing the metal to study, at an initial concentration established and adding 10 mL culture of 24 h of each strain, with a biomass concentration of 1 g/L. Target used 100 mL of metal solution without bacteria. Samples were analyzed in duplicate for each strain. Is used nephelometer of McFarland to estimate the number of cells/mL [40, 42]. The conditions established are: pH between 4 and 5, if the metal precipitates at neutral pH, 37 °C and 100 rpm agitation [44, 45]. To read the metal concentration is done by atomic absorption spectrophotometry, taking 5 mL sample every 15 minutes and prepared as described by [23, 46] and the concentration is calculated from the calibration curve prepared with standard solution of each metal studied. The detection limits can be 0.02 mg/L, analyzing in duplicate.

3.6. Affecting parameters

Within the anaerobic and aerobic environment, various important parameters affect the rates of the different steps of the process, i.e. pH and alkalinity, temperature, and hydraulic retention times.

Each group of microorganisms has a different optimum pH range. Methanogenic bacteria are extremely sensitive to pH with an optimum between 6.5 and 7.2 pH, alkalinity and volatile acids/alkalinity ratio. The fermentative microorganisms are somewhat less sensitive and can function in a wider range of pH between 4.0 and 8.5. at a low pH the main products are acetic and butyric acid, while at a pH of 8.0 mainly acetic and propionic acid are produced.

The temperature has an important effect on the physicochemical properties of the components found in the digestion substrate. It also influences the growth rate and metabolism of microorganisms and hence the population dynamics in the anaerobic reactor. Acetotrophic methanogens are one of the most sensitive groups to increasing temperatures. The degradation of propionate and butyrate is also sensitive to temperatures above 70 °C. The temperature has moreover a significant effect on the partial pressure of H_2 in digesters, hence influencing the kinetics of the syntrophic metabolism. Thermodynamics show that endergonic reactions (under standard conditions), for instance the breakdown of propionate into acetate, CO_2, H_2, would become energetically more favourable at higher temperature, while reactions which are exergonic (e.g. hydrogenotrophic methanogenesis) are less favoured at higher temperatures.

The solids retention time (SRT) is the average time the solids spend in the digester, whereas the hydraulic retention time (HRT) is the average time the liquid sludge is held in the digester. The subsequent steps of the digestion process are directly related to the SRT. A decrease in the SRT decreases the extent of the reactions and viceversa. Each time sludge is withdrawn, a fraction of the bacterial population is removed thus implying that the cell growth must at least compensate the cell removal to ensure steady state and avoid process failure [28].

3.7. McFarland nephelometer

The McFarland nephelometer was described in 1907 by J. McFarland as an instrument for estimating the number of bacteria in suspensions used for calculating the bacterial opsonic index and for vaccine preparation.

Another important factor is known that cells per milliliter are taken at a given time, and a known way is through the % of transmittance and that can be determined by the technique of McFarland nephelometer, which is described below.

McFarland Nephelometer Standards

a. Set up 10 test tubes or vials of equal size and quality: Use new hoses washed and completely dry.
b. Prepare H_2SO_4 1% chemically pure.
c. Prepare an aqueous solution of barium chloride, 1% chemically pure.
d. Add to the tubes the designated amounts of the two solutions as shown in Table 2 for a total of 10 mL/tube.
e. Close the tubes or vials. The suspension of barium sulphate corresponding to an approximately homogeneous precipitate of density of the cells per milliliter in the standard variable, as shown in the Table 2.
f. In the Figure 3(a) shows the % of transmittance against the number. From the tube which can be removed if there is not a spectrophotometer to read the transmittance, and in Figure 3(b) shows the number of cells that are depending on number.

Tube's Number	1	2	3	4	5	6	7	8	9	10
Barium chloride (mL)	0.1	0.2	0.3	0.4	0.5	0.6	0.7	0.8	0.9	1
Sulfuric Acid (mL)	9.9	9.8	9.7	9.6	9.5	9.4	9.3	9.2	9.1	9
Aprox. Cell density $(X10^8/mL)$	3	6	9	12	15	18	21	24	27	30

Table 2. McFarland Nephelometer Standards

Population density is monitored by taking readings of % of transmittance, which compared to the McFarland nephelometer; transmittance readings should be less than 10% in order to maintain the population density. This technique is used to know the time and turbidity of most practical way to achieve the desired amount of biomass for biosorption experiments batch system and continuous [39].

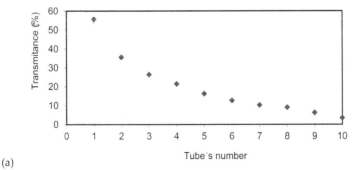

(a)

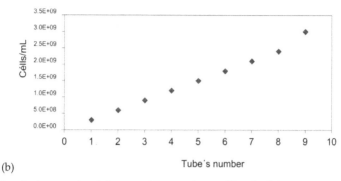

(b)   Tube's number

Figure 3. Standard curves of nephelometer with transmittance (%) and cell density

4. System in batch and continuous culture

The construction of bioreactors is based on a simple principle: make the pollutants are converted into the substrate (food) of microorganisms, and that these, while feeding and increases its population, decontaminated water. For the construction of a bioreactor is necessary to know the type of microorganisms with which they are going to work and as well as the growth curve characteristic of them [5].

The key factor of a bioreactor is to maintain microorganisms in the growth stage most of the time as possible, i.e. keep the microbial population to its maximum level, to optimize the efficiency of the degradation processes. This is achieved by controlling the environmental conditions (temperature, pH, aeration and nutrient availability) and the flows in and out, so never lack food and do not reach the death phase or endogenous [2, 5].

The teams that are made homogeneous reactions can be of three general types: discontinuous (batch), continuous flow steady and unsteady flow semicontinuous.

Batch reactors are simple to operate and industrially used when small amounts are to treat substance. Continuous reactors are ideal for industrial purposes be treated when large quantities of substance and can achieve good control of product quality. Semicontinuous reactors are more flexible systems, but more difficult to analyze and operate than previous; in them the reaction rate can be controlled with a good strategy at the dosage of the reactants [48].

In a perfect batch reactor there is no entry or exit of reactant. It is further assumed that the reactor is well stirred, i.e. that the composition is the same at all points of the reactor for a given time instant. Since the input and output are zero the material balance is:

$$\begin{pmatrix} Disappearance\ of \\ reactant\ by \\ chemical\ reaction \end{pmatrix} = -\begin{pmatrix} Reactant \\ accumulation\ in\ the \\ control\ volume \end{pmatrix}$$

All points have the same composition; the volume control to perform the balance is the entire reactor. Evaluating the terms:

$$r_A V = -\frac{dN_A}{dt} \qquad (11)$$

And given that: $N_A = N_{A0}(1 - X_A)$ results:

$$r_A V = N_{A0}\frac{dX_A}{dt} \qquad (12)$$

Integrating gives the equation for the design for the batch reactor:

$$t = N_{Ao}\int_o^{X_A} \frac{dX_A}{r_A V} \qquad (13)$$

If the reaction volume remains constant may be expressed in function of the concentration of reagent $C_A = N_A / V$

$$t = C_{Ao}\int_o^{X_A} \frac{dX_A}{r_A} = -\int_{C_{Ao}}^{C_A} \frac{dC_A}{r_A} \qquad (14)$$

This intermittent or batch reactor is characterized by the variation in the reaction's degree and the properties of the reaction mixture with the lapse of time [49]

A batch reactor has no inflow or outflow reagents of the reaction products while being performed. In almost every batch reactors, the longer that a reactant in the reactor, most of it becomes product to reach equilibrium is exhausted or the reagent [50].

The reactor of the continuous flow type, in which the degree of reaction can vary with respect to the position in the reactor, but not a function of time. Therefore, one of the classifications of the reactors is based on the operation method [49].

Normally, the conversion increases with time that the reagents remain in the reactor. In the case of continuous flow systems, this time usually increases with increasing reactor volume; therefore, the conversion X is a function of reactor volume V [50].

The tubular reactor plug flow (RTFP) is characterized in that the flow is directed, without any element of the exceeding or being mixed with any other element located before or after that, i.e. no mixing in the flow direction (axial direction). As a result, all fluid elements have the same residence time within the reactor [48].

As fluid composition varies along the reactor, material balance must be performed in a differential volume element transverse to the direction of flow.

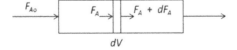

dV

Inlet = Outlet + Disappearance of reaction

$$F_A = F_A + dF_A + r_A dV$$

Given that $dF_A = d[F_{Ao}(1-X_A)] = -F_{Ao}dX_A$ by substitution is

$$F_{Ao}dX_A = r_A dV \tag{15}$$

That integrated is

$$\int_0^V \frac{dV}{F_{Ao}} = \int_0^{X_{Af}} \frac{dX_A}{r_A} \tag{16}$$

Or:

$$\tau = \frac{V}{V_o} = C_{Ao}\int_0^{X_{Af}} \frac{dX_A}{r_A} \tag{17}$$

In some cases it is convenient to divide the output current of a plug flow reactor by returning part of it to the reactor inlet.

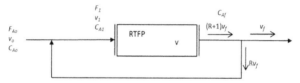

The recirculation ratio is defined:

$$R = \frac{\text{flow which is recycled}}{\text{flow out}} \tag{18}$$

Raising the design equation for the reactor (within the recycle loop) without expansion

$$\frac{V}{V_1} = -\int_{C_{A1}}^{C_{Af}} \frac{dC_A}{r_A} \tag{19}$$

If its considers that there is no expansion or contraction in the reactor, raising the junction of the inlet and the recirculation $V_1 = (R + 1) V_o$ and furthermore $C_{A1} = (C_{Ao} + C_{Af}) / (R + 1)$, therefore the equation of reactor design is:

$$\tau - \frac{V}{V_o} - -(R+1)\int_{\left(\frac{C_{Ao}+RC_{Af}}{R+1}\right)}^{C_{Af}} \frac{dC_A}{r_A} \tag{20}$$

Another classification relates to the shape. If a laboratory vessel is equipped with a stirrer efficient, composition and temperature of the reaction mass will tend to be equal in all areas of the reactor. A container in which there is uniformity of properties is called a stirred tank reactor (or well mixed) or STR [48].

4.1. Anaerobic continuous studies

Generally, the design rules of biological reactors are all based on high removal efficiency of degradable organic matter. Consequently, if the substrate composition and strength of a wastewater are known, a basic design of a high-rate anaerobic system can be established. According to [52], the main design criteria of UASB reactors are, among others: applicable organic load, upflow velocity, three-phase separator, and influent distribution system.

The UASB reactors are generally designed based on the organic volumetric load (OVL) (kgCOD/m³-day) that is defined as follows:

$$OVL = (QSo)/V \tag{21}$$

where Q: influent flow rate (m^3/day), So: influent COO ($kgCOD/m^3$) , and V: volume of reactor (m^3), From Equation (21) the volume of the reactor, V, can be obtained:

$$V = QSo/OVL \tag{22}$$

For most industrial wastewaters, the OVL (based on degradable COD) is the critical factor for the reactor volume. Its value depends on the quantity and quality of the granular sludge; the nature, type, and concentration of the pollutants; the temperature; the required treatment efficiency and the desired safety regarding peak loads [16, 53].

4.2. Studies of biosorption of heavy metals with aerobic bacteria using biomass support in batch system

For studies of biosorption of heavy metals using aerobic bacteria and a support for the biomass, using 500 mL Erlenmeyer flasks, which are placed 5 g of the support for the immobilization of selected biomass, 90 ml of a solution containing the metal study established at an initial concentration, 10 ml of biomass with a density of 1 g/L and as target, 100 mL of metal with 5 g of support material. The flasks were plugged with a cotton swab having aeration, then is placed in an incubator with shaking at 100 rpm and temperature established for mesophilic bacteria to 45 °C. Samples are taken at set times to analyze the concentration of metals by atomic absorption. The conditions are the same for the studies using only bacteria without carrier material. All experiments were performed in duplicate and the efficiency of biosorption (E) is calculated using the equation:

$$E = \left(\frac{C_o - C_f}{C_o} \right) * 100 \tag{23}$$

Where: C_o y C_f initial and final amounts of the metal (mg/L).

4.3. Studies of heavy metal biosorption in a continuous system with aerobic biomass using biomass support

Biosorption is a rapid phenomenon of passive metal sequestration by the non-growing biomass. To carry out the studies of metal biosorption in a continuous system is conditioned first a reactor column, which has side ports for sampling. The reactor is packed with carrier material of biomass with a particle size between 1 and 6 mm, to avoid clogging. Both mineral medium such as air are fed through the bottom of the reactor to promote the growth of bacteria and the pH is controlled if the metal to be studied could precipitate at neutral pH. The mineral medium is inoculated with 10% biomass that develop in this medium for the time given the growth kinetics and the reactor is kept in recirculation until the development of biomass (1 g/L) and that adheres to support material. Biomass concentration was estimated by measuring the percent of transmittance to an optical density at 600 nm (Spectronic 20D⁺) and for the amount of biomass produced in cells/mL was determined using the Table 2 of McFarland nephelometer, described above.

When produced in the reactor is 1 g/L of biomass and this is immobilized in the zeolite, are set constant conditions of operation of the reactor as: air flow 10 times the feed flow of the contaminated medium, hydraulic retention time (HRT) of a day and ambient temperature of 30 °C. The tests are performed to set conditions of initial concentrations of metals and pH set, and takes days to the input samples at different heights of the reactor and output to meet the metal concentration, biomass is recycled to make more time contact between the bacteria and the metal being studied. It can be perform a second and third experimental run at different initial concentrations of feeding and at the same pH, maintaining the feed stream and recirculation same. Other studies may be changing the pH, keeping other conditions constant.

In the experimental runs carried out is analyzed for pH, metal concentration by atomic absorption and to determine the concentration of cells/mL of biomass, measures the percentage of transmittance in the spectronic 20D⁺ and compared by the technique of Nephelometer of Mc Farland. The support used is analyzed by the technique of sludge digestion, are performed analyzes of biomass produced per day and chemical oxygen demand (COD). At the end of the experiments are performed technical analyses of the medium used X-ray Diffraction (XRD), scanning electron microscopy (SEM) and Energy Dispersive Spectroscopy X-ray (EDS) at different column heights to see if deposited on the support certain amount of heavy metals or all was absorbed by the biomass.

4.4. Support the immobilization of biomass

Immobilization of cells as biocatalysts is almost as common as enzyme immobilization. Immobilization is the restrict ion of cell mobility within a defined space. Immobilized cell cultures have the following potential advantages over suspension cultures.

- Immobilization provides high cell concentrations.
- Immobilization provides cell reuse and eliminates the costly processes of cell recovery and cell recycle.
- Immobilization eliminates cell washout problems at high dilution rates.
- The combination of high cell concentrations and high flow rates (no washout restrictions) allows high volumetric productivities.

- Immobilization may also provide favorable microenvironmental conditions (i.e., cell-cell contact, nutrient-product gradients, pH gradients) for cells, resulting in better performance of the biocatalyst» (e.g., higher product yields and rates).
- In some cases, immobilization improves genetic stability.
- For some cells, protection against shear damage is important.

The major limitation on immobilization is that the product of interest should be excreted by the cells. A further complication is that immobilization often leads to systems for which diffusional limitations are important. In such cases the control of microenvironmental conditions is difficult, owing to the resulting heterogeneity in the system. With living cells, growth and gas evolution present significant problems in some systems and can lead to significant mechanical disruption of the immobilizing matrix.

The primary advantage of immobilized cells over immobilized enzymes is that immobilized cells can perform multistep, cofactor-requiring, biosynthetic reactions that are not practical using purified enzyme preparations.

Adsorption of cells on inert support surfaces has been widely used for cell immobilization. The major advantage of immobilization by adsorption is direct contact between nutrient and support materials. High cell loadings can be obtained using microporous support materials. However, porous support materials may cause intraparticle pore diffusion limitations at high cell densities, as is also the case with polymer-entrapped cell systems. Also, the control of microenvironmental conditions is a problem with porous support materials. A ratio of pore to cell diameter of 4 to 5 is recommended for the immobilization of cells onto the inner surface of porous support particles. At small pore sizes. Accessibility of the nutrient into inner surfaces of pores may be the limiting factor, whereas at large pore sizes the specific surface area may be the limiting factor. Therefore, there may be an optimal pore size, resulting in the maximum rate of bioconversion.

Adsorption capacity and strength of binding are the two major factors that affect the selection of a suitable support material. Adsorption capacity varies between 2 mg/g (porous silica) and 250 mg/g (wood chips). Porous glass carriers provide adsorption capacities (10^8 to 10^9 cells/g) that are less than or comparable to those of gel-entrapped cell concentrations (10^9 to 10^{11} cells/mL). The binding forces between the cell and support surfaces may vary, depending on the surface properties of the support material and the type of cells. Electrostatic forces are dominant when positively charged support surfaces (ion exchange resins, gelatin) are used. Cells also adhere on negatively charged surfaces by covalent binding or H bonding. The adsorption of cells on neutral polymer support surfaces may be mediated by chemical bonding, such as covalent bonding, H bonds, or van der Waals forces. Some specific chelating agents may be used to develop stronger cell-surface interact ions. Among the support materials used for cell adsorption are porous glass, porous silica, alumina, ceramics, gelatin, chitosan, activated carbon, wood chips, polypropylene ion-exchange resins (DEAE-Sephadex, CMC-), and Sepharose [54].

Various reactor configurations can be used for immobilized cell systems. Since the support matrices used for cell immobilization are often mechanically fragile, bioreactors with low hydrodynamic shear, such as packed-column, fluidized-bed, or airlift reactors, are preferred. Mechanically agitated fermenters can be used for some immobilized-cell systems if the support matrix is strong and durable. Any of these reactors can usually be operated in a perfusion mode by passing nutrient solution through a column of immobilized cells [54].

Since the design of reactors for the removal of heavy metals from liquid effluent must consider optimum contact between these and the biomass, it has been considered the use of different types of support for the immobilization of the biomass with the aim of achieving greater efficiency in removing heavy metals. This achieves prevent biosorbent is removed from the reactor in the output current and at the same time is obtained a greater mechanical stability thereby reducing the shear stresses that could damage the structure of the microorganism which affects removal efficiency heavy metals [53].

Living biomass immobilized, must first take the form of biofilm on supports prepared from a variety of inert materials. One of the materials that have been studied as biomass support is activated carbon by porosity and high surface area, besides being an abundant product is obtained as a byproduct of the production of oil from coconut, olive and processing sugarcane [53]. Other materials have been used as biomass support such as silica, polyacrylamide gel and polyurethane include agar, cellulose, alginates, polyacrylamides, the silica gel, sand, textile fibers, calcium alginate, polysulfone, glutaraldehyde and other organic compounds, and have been used for removing heavy metals [55,56].

There are other materials that could be used for biomass carriers; such as the natural zeolites are known important industrial applications due to its high affinity for water and that the cavities only allow passage of molecules of a certain size. Have been used as additives in animal feed, such as soil improvers in agriculture due to increased nitrogen retention and soil moisture, and as catalysts in industrial processes of refining, petrochemicals and fine chemicals [57].

4.4.1. Activated charcoal

The name of activated charcoal is applied to a series of artificially prepared porous carbons to exhibit a high degree of porosity and a high inner surface. These characteristics are responsible for their adsorptive properties, which are used widely in many applications in gas phase and liquid phase. Chemically it is composed of carbon, oxygen, hydrogen and ash. The activated carbon adsorbent is a very versatile, because the size and distribution of pores in the carbonaceous structure can be controlled to meet the current and future technology. The pore sizes ranging from smaller called micropores (2.0 nm) until the mesopores (2 - 50 nm) and macropore (<50 nm). It should be borne in mind that most adsorption occurs in the micropores (greater than 90% of the surface area) the mesopores and macropores are extremely important because in the activated charcoal are those which facilitate access of the species will adsorb to the interior of the particle and of the micropores [58].

One of the materials that have been studied as biomass support is activated charcoal. Its high porosity and high surface area activated charcoal make it an ideal material to be carried out the process of adsorption of heavy metals. Another reason why activated charcoal is used for the adsorption is its low cost, since it is an abundant product is obtained as a byproduct of the production of oil from coconut, olive and processing of sugar cane [55].

4.4.2. Glutaraldehyde

Microbial cells can be immobilized by cross-linking between cells, using bi or multifunctional reagents as glutaraldehyde or toluene di isocyanate

Glutaraldehyde is a colorless liquid with a pungent odor used to sterilize medical and dental equipment is also used in water treatment industry and as a chemical preservative. However, it is toxic and can cause severe eye irritation, nose, throat and lungs, along with headaches, drowsiness and vomiting. Glutaraldehyde monomer can polymerize by aldol condensation, giving poliglutaraldehído alpha, beta unsaturated reaction typically occurs at alkaline pH [59].

4.4.3. Silica

The mechanism involved is based on the formation of covalent bonds between the inorganic support (silica) and cells in the presence of crosslinking agents. A said joint is needed for the modification of the support surface. The reaction requires the introduction of reactive organic groups on the silica surface for the attachment of cells to the support. As coupling agent generally used aminopropyl triethoxy silane; this organic functional group condenses with hydroxyl groups of the silica and the group as a result becomes available for covalent bond formation on the surface. Covalent bonds can also be established by treating the silica surface with glutaraldehyde or isocyanate.

The advantage of this method is that the support can be generated without the limitations on physical and chemical conditions imposed by the biocatalyst, which can be optimized by the characteristics of mechanical stability, porosity, strength of the support, etc.

When you want to form covalent bonds between the substrate and cells, the problem is how to promote adhesion of cells to relatively large surface without damaging its stability and resistance to washing. The support may have pores of greater diameter than the cell to allow the latter to penetrate the internal surfaces. Porous supports are used which are embedded by immersion in cell suspensions [60].

Another important matrix being used for immobilization for metal removal is silica. Silica-immobilized preparations offer advantage in terms of reusability and stability. The silica immobilized product is mechanically strong and exhibits excellent flow characteristics [68]. A silica immobilized algal preparation AlgaSORBR (Bio-Recovery Systems, Inc., Las Cruces, NM 80003, USA) which is being used commercially retains approximately 90% of the original metal uptake efficiency even after prolonged use (> 18 months) [57].

4.4.4. Polyacrylamide gel

Polyacrylamide gels are formed by polymerization of acrylamide by the action of a crosslinked agent, is chemically inert, uniform properties, able to be prepared quickly and reproducibly. Thus, in addition, transparent gels with mechanical stability, water insoluble, relatively non-ionic and allow good visualization of the bands for a long time. Also has the advantage that by varying the concentration of polymers can be modified in a controlled manner the pore size, there sometimes is used in diagnosis least because of their neurotoxocidad [59].

Whole cell immobilization within a polyacrylamide gel also provides a useful laboratory scale system and has been used to biosorb and recover a number of heavy metal(s). Good results have been obtained in the case of polyacrylamide immobilized cells of Citrobacter where a very high removal of uranium, cadmium and lead was observed from solutions supplemented with glycerol $-2PO_4$. Rhizopus arrhizus biomass immobilized on polyacrylamide gel was effective in almost completely removing Cu^{2+}, Co^{2+} and Cd^{2+} from synthetic metal solution [26, 55].

4.4.5. Polyurethane

Inert materials such as polyurethane, impregnated with a suitable culture medium provided a homogeneous aerobic condition in the fermenter and impurities do not contribute to the final product. An additional advantage of using inert supports is the easy recovery of the product of interest, ease of performing balances because all nutrient concentrations in the middle of production are known, so one can study the effect of a given component of medium [60].

Recent studies have shown the superiority of polyurethane and polysulfone as immobilization support in comparison to polyacrylamide and alginate matrices. It has been reported a novel polyurathane gel bead fabrication technique for immobilizing Pseudomonas aeruginosa CSU. Preliminary studies conducted by them revealed that the P. aeruginosa CSU biomass immobilized within the polyurethane gel beads were effective in the removal of hexavalent uranium from low concentration acidic waters. Other authors have been immobilized phormidium laminosum on polysulfone and epoxy resins. They were successful in reusing the polysulfone immobilized biomass for ten consecutive biosorption/desorption cycles without apparent loss of efficiency after reconditioning it with 0.1 M NaOH. Immobilization of Citrobacter biomass in polysulfone matrix increased its metal loading capacity for lead, cadmium and zinc metals [57].

4.4.6. Alginate

There are many studies on the composition of alginate and its advantages to cell immobilization. Because the chemical composition of this polymer and as a consequence that the same reactions can be obtained in reaction, alginate gels are recommended for cells sensitive to environmental conditions. Recent studies of the diffusional characteristics of the immobilized system, have improved our understanding of the environment surrounding the immobilized cells, optimize protocols and improve the stability of alginate gels [60].

One of the matrices that have been used in metal recovery by both viable and non-viable cells is the entrapment in the matrix of insoluble Ca-alginate. Fluidized beds of Ca-entrapped cells of Chlorella vulgaris and Spirulina platensis were successfully used to recover gold from a simulated gold-bearing process solution containing $AuCl_4$, $CuCl_2$, $FeCl_2$ and $ZnCl_2$. The Ca-alginate immobilized cells of Chlorella salina also showed greater binding of cobalt, zinc and manganese than the free cells. Rhizopus arrhizus entrapped on alginate beads was successfully used for the removal of uranium over multiple biosorption and desorption cycles. Accumulation was also dependent on cell density in alginate beads with greater uptake of cobalt at the highest cell densities [57].

4.4.7. Natural zeolite

Zeolites are crystalline aluminosilicates, three-dimensional, microporous, based on framework structure with a rigid anion, with well-defined channels and cavities. These cavities contain exchangeable metal cations (Na^+, K^+, etc.) And can also retain removable and replaceable guest molecules (water in natural zeolites). To date about 40 have been characterized structures of natural zeolites and have developed more than 130 synthetic structures. The most important natural zeolites are analcime, chabazite, clinoptilolite, erionite, ferrierite, heulandite, laumontite, and phillipsite mordonita [58].

Zeolites are composed of aluminum, silicon, sodium, hydrogen and oxygen. The crystal structure is based on the three network addresses with SiO_4 tetrahedral shaped with four oxygens shared with adjacent tetrahedra. The physical properties unique aspects provide for a wide variety of practical applications. Figure 4 shows the basic structure of the zeolite tetrahedral [63].

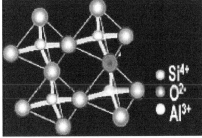

Figure 4. Basic tetrahedral structure of the zeolites.

The physical properties of the zeolite are that they possess features bright, hardness and wear resistance. Applications of natural zeolites make use of one or more of its chemical properties, usually including adsorption, ion exchange and catalysis. These properties are a function of the crystal structure of each species, cationic structure and composition [63].

Clinoptilolite is from the zeolite minerals are best known for its uses and applications. It is a natural zeolite formed from volcanic ash in lakes and marine waters millions of years ago. Clinoptilolite, is the most studied and is considered more useful, since it is known as an adsorbent of certain toxic gases such as hydrogen sulfide and sulfur dioxide. In fact few countries that have had deposits in operation, including: Japan, Italy, USA, Russia, Hungary, Bulgaria, Cuba, Yugoslavia and Mexico [39]. Recognizes the capacity of the zeolites natural adsorb heavy metals and other contaminants from water. In certain cases, it requires a pretreatment of the zeolite to modify or improve its adsorption properties [62].

5. Conclusion

This chapter has provided the results of research of aerobic and anaerobic biomass in the batch and continuous system using a support for the biomass: silica, polyacrylamide gel, polyurethane, calcium alginate, glutaraldehyde, charcoal and zeolites. The use of support for the biomass increases the development of the microorganisms. Also describes the affecting parameters: time, pH, temperature, HRT, toxicity and stirring speed. Also in this chapter describe the techniques for determination of parameters for anaerobic such as, chemical oxygen demand (COD), alkalinity, methane production, total solids (TSS), volatile solids (VSS), volatile fatty acids (VFA) concentrations, and for the aerobics, the biomass concentration using % of transmittance, McFarland nephelometer, isolation, macroscopic and microscopic characterization, growth kinetics, in batch and continuous system.

Author details

Onofre Monge Amaya, María Teresa Certucha Barragán and Francisco Javier Almendariz Tapia
University of Sonora, Department of Chemistry and Metallurgy. Hermosillo, Sonora. México

Acknowledgement

This work was made possible through support provided by the University of Sonora, through the Department of Chemical Engineering and Metallurgy, and Engineering Division. The authors would like to thank: The National Council for Science and Technology (CONACyT), well as students Gisel Figueroa, Gonzalo Figueroa, Guadalupe López, Karla Hernandez, Hiram Bañuelos, Carlos Jaramillo, Luis Carlos Platt, Axel Valenzuela, Glenda Duarte.

6. References

[1]   Madigan M.T., Martinko J. M., Parker J. (2004) Brock Biología de los microorganismos, ED. Prentice Hall. 647 p.

[2]   Eweis J. B., Ergas S. J., Chang D. P., Schroeder E. D., Tejero M. I., Amieva del Val J.J. (1999) Principios de biorrecuperación (Bioremediation), Primera Edición, Mac Graw Hall Interamericana. 4, 12, 21, 94.

[3]   Atlas. M.R. y Bartha R. (2002) Ecología microbiana y microbiología ambiental. 4a Ed.
      Pearson educación. Madrid, España. pp. 3-24.
[4]   Steciow M. (2005) Microbiología Ambiental
      http:www.cricyt.edu.ar/enciclopedia/términos/Biorremed. Htm
[5]   Fontúrbel R. F. e Ibáñez N.C. (2004). Empleo del Metabolismo Microbiano para la
      Descontaminación de Aguas. Biología y Ciencias de la Salud. No. 17.
[6]   Saigí F. y López A. (2004) Las ciencias de la vida y la biotecnología en la nueva
      sociedad del conocimiento. La base de la nueva economía.
      http://www.uoc.edu/dt/esp/saigi1104.pdf. Agosto-2008.
[7]   Basso, M.C., Cerella, E.G.,Cukierman, A.L (2002) Empleo de Algas Marinas para la
      Biosorción de Metales Pesados de Aguas Contaminadas. Avances en Energías
      Renovables y Medio Ambiente. Argentina. 6:1
[8]   Cotoras T. D. (2003) Bacterias para descontaminar aguas residuales de la minería y la
      industria. Facultad de Ciencias Químicas y Farmacéuticas. Universidad de Chile.
      Induambiente 13(14): 27-29.
      http://www.estrucplan.com.ar/Secciones/Noticias/Noticia.asp. Abril 2012
[9]   Mijares M. A. P. (2003) Aislamiento y Caracterización de Bacterias con Capacidad para
      Inmovilizar Cadmio. Tesis Licenciatura. Biología con área en Biotecnología.
      Universidad de las Américas, Puebla. Capítulo dos.
[10] Cervantes Carlos (2006) Las Relaciones de las Bacterias y los Metales. Carisma de la
      Ciencia 3: 1-4.
[11] Rivas B.G.A., Gutierréz, S., Merino F. (2004) Bioremoción de Metales Pesados en
      Solución por *Pseudomonas fluorescens* M1A-45 Aisladas de Ambientes Minero. Segunda
      Semana de Ciencia y Tecnología.
[12] Acosta I., Moctezuma-Zárate M. G., Cárdenas J. F. y Gutiérrez C. (2007) Bioadsorción de
      Cadmio (II) en Solución Acuosa por Biomasa Fúngicas. Información Tecnológica. 18:1:
      9-14.
[13] Ilhan S., Nourbakhsh M.N., Kilicarslan S., Ozdag H. (2004) Removal of Chromium,
      Lead and Copper Ions From Industrial Waste Waters by *Sthaphlococcus saprophyticus*
      Turkish Electronic Journal of Biotechnology. 2, 50-57.
[14] Lasat M. M. (2002) Phytoextraction of toxic metals: A review of biological mechanisms.
      J. Environ. Qual. 31,109-120.
[15] Monge, O., Valenzuela, J., Acedo, E., Certucha, M., & Almendáriz, J. (2008) Biosorción
      de cobre en sistema por lote y continuo con bacterias aerobias inmovilizadas en zeolita
      natural (clinoptilolita). Rev. Int. Contam. Ambient. 24 (3) 107-115.
[16] Certucha-Barragán M.T, Acedo-Félix E,   Monge-Amaya O, Valenzuela-García J.L,
      Almendariz-TapiaF.J, Leal-Cruz A.L (2009) Copper Bioaccumulation in an Upflow
      Anaerobic Sludge Blanket (UASB) Reactor Chemical Speciation and Bioavailability.
      21:3: 161-164.
[17] Certucha-Barragán M. T, Acedo-Félix F, Almendariz-Tapia F. J, Pérez-Moreno R,
      Monge-Amaya, Valenzuela Garcia J. L, and O. Monge-Amaya (2011) Iron Influence on
      Copper Biosorption Using Anaerobic Sludge and its Microstructural Characterization,
      Mineral Processing & Extractive Metall. Rev., 32: 1 - 8.

[18] Topalián M.L., Castané P.M, Rovedatti, M.G. and Salibián A. (1999) Principal Component Analysis of Dissolved Heavy Metals in Water of the Reconquista River (Buenos Aires, Argentina). Bull. Environ. Contam. Toxicol 63: 484 – 490.

[19] Karri S, Sierra-Alvarez, R and Field J.A. (2006) Toxicity of Copper to Acetoclastic and Hydrogenotrophic Activities of Metanogens and Sulfate Reducers in Anaerobic Sluge. Chemosphere, 62: 121 – 127.

[20] Upadhyay, A.K., Gupta, K.K., Sircar, J.K., Deb, M.K. and Mundhara, G.L. (2006) Heavy Metals in Freshly Deposited Sediments of the Rivers Subernarenkha, India: an Example of Lithogenic and Anthtopogenic Effects Environ. Geol., 50:397–403.

[21] Elder, F.J. (1988) Metal Biogeochemistry in Surface-Water Systems A Review of Principles and Concepts, U. S. Geologycal Survey Circular 1013, U.S.A

[22] Alloway, B.J. and Ayres, D.C. (1993) Chemical Principles of Environmental Pollution. Blackie Academic & Professional. New York, USA. pp. 291.

[23] Gomez-Alvarez et al., 2007 Gómez–Álvarez, A., Valenzuela-García, J. L., Aguayo-Salinas, S., Meza-Figueroa, D., Ramirez-Hernández, J., Ochoa-Ortega, G. (2007) Chemical partitioning of sediment contamination by heavy metals in the San Pedro River, Sonora, Mexico. Chemical Speciation and Bioavailability. 19, 25-36.

[24] Gómez-Álvarez, A. (2001). Evaluación de la calidad física y química del agua y sedimento del Río San Pedro, Sonora, México, Durante el Periodo 1997-1999. Tesis de Maestría en Ciencias de la Ingeniería. Universidad de Sonora

[25] Henry J. G. y Heinke W.G. (1999) Ingeniería Ambiental. Ed Prentice Hall. pp. 195, 196, 431.

[26] Speece R.E. (1996) Amaerobic Biotchnology.Archae Press.pp.24-63

[27] Burak Demirel, Paul Scher ,Orhan Yenigun, and Turgut Onay (2010) Production of Methane and Hydrogen from Biomass through Conventional and High-Rate Anaerobic Digestion Processes Critical Reviews in Environmental Science and Technology. 40:116–146.

[28] Lise Appels , Jan Baeyens , Jan Degre`ve , Raf Dewil (2008) Principles and Potential of the Anaerobic Digestion of Waste-Activated Sludge Progress in Energy and Combustion Science. 34: 755–781.

[29] A.c. van Haanadel 2006 Advanced Biological Treatment Processes for Industrial Wastewaters Principles and Applications.In: Francisco 1.Cervantes, Spyros G. Pavlostathis and Adrianus C. van Haandel: in IWA.pp. 67-70.

[30] Gujer, W. and Zehnder, AJ. (1983) Conversion Processes in Anaerobic Digestion Wat.Sci.Techn. 15,127.

[31] Chiu-Yue LIinm and Chin-Chao Chen (1999) Effect of Heavy Metals on the Methanogenic UASB Granule Wat. Res. 33, 2, pp. 409-416.

[32] Henzen, M. and Harremoes, P. (1983) Anaerobic Treatment of Waste Water in Fixed Film Reactors – a Literature Review Wat.Sci. Techn. 15, l.

[33] Sandoval C, Carreño M. De, Castillo E F, Vergara M. (2007) Sludge Anaerobic Microbiologic Characterization Used in the Organic Fraction of Urban Solid Waste Treatment Scientia et Technica Año XIII, 35:509-514.

[34] APHA. *Standard Methods for the Examination of Water and Wastewater* (1995) Washington DC, USA: American Public Health Association/American Water Works Association/Water Environment Federation.

[35] Bouvier J.C., Steyer J.P. and Delgenes J.P. (2002) On-line titrimetric sensor for the control of VFA and/or alkalinity in anaerobic digestion processes treating industrial vinasses, VII Latin American Workshop and Symposium on Anaerobic Digestion, Merida, Mexico, October 23-25/2001, 65-68.

[36] Almendariz F. J. (2005) Tratamiento de Sosas Gastadas en un Reactor de Lecho Expandido de Lodo Granular Anaerobio Tesis Doctoral. Universidad Autónoma Metropolitana. México, D.F.

[37] Alves M., Cavaleiro A.J., Ferreira E.C., Amaral A.L, Mota M., da Motta M., Vivier H. and Pons M-N. (2000) Characterisation by image analysis of anaerobic sludge under shock conditions. Water Science and Technology Vol 41 No IWA Publishing 12. 207–214.

[38] Laguna A., Ouattara A., González R. O., Barón O., Fama G., El Mamouni R., Guiot S., Monroy O. y Macarie H. (1999) A Simple and Low Cost Technique for Determining the Granulometry of Upflow Anaerobic Sludge Blanket Reactor Wat. Sci. Technol 40 (8). 1-8.

[39] Onofre Monge, Leobardo Valenzuela y Evelia Acedo (2010) *Biosorción de Cobre con Bacterias Aerobias Inmovilizadas en Zeolita. Aislamiento, Caracterización, Selección de Microorganismos y Cinética de Biosorción.* Lambert Academic Publishing.

[40] Aquiahuatl R. M.A. y Pérez Ch. M.L. (2004) *Manual de prácticas del laboratorio de microbiología general.* UAM. México. Pp. 37-49.

[41] Koneman E. W., Allen S., Janda W., Schrenberger P., Winn W. (1999) Diagnóstico microbiológico. Médica Panamericana. México. pp. 1-349; 1258-1357.

[42] Mac Faddin J. F. (2007) Pruebas bioquímicas para la identificación de bacterias de importancia clínica. Médica Panamericana. Pp. 94-421.

[43] Abuzer Çelekli a, Mehmet Yavuzatmacab, Hüseyin Bozkurtc (2010). An ecofriendly process: predictive modelling of copper adsorption from aqueous solution on *Spirulina platensis.* Journal of Hazardous Materials. Vol.173, 123–129.

[44] Monge A. O. (2003) Biorrestauración de suelos contaminados por cianuro, utilizando Bacillus sp. Tesis de Maestría en Ciencias de la Ingeniería. Universidad de Sonora.

[45] Duarte S. D. (1997). Evaluación de la capacidad de biosorción de Ni(II) por Thiobacillus Ferrooxidans. Tesis de Licenciatura en Ingeniería Química. Universidad de Sonora.

[46] Oliveira M. J. (2003). Estudio de la biosorción de cobre por perlas de alginate de calcio. Tesis de Licenciatura. Universidad Nacional Mayor de San Marcos. Perú.

[47] Agemian H. y Chau A. S. Y. (1975) An atomic absorption method for the determination of 20 elements in lake sediments after acid digestion. Analytica Chemica Acta 80:61-66.

[48] Borzacconi L. y López I. (2003) Cinética de ingeniería de reacciones. Notas de curso. Facultad de Ingeniería. http://www.fing.edu.uy/iq/reactores/cursos/reactores1.pdf consulta: Abril-2012.

[49] Smith J. M. (1991) Ingeniería de la Cinética Química. ED. Continental. México D. F. pp. 369-377.

[50] Fogler, H. S., Escalona G. R. L. y Ramírez S. J. F. (2001) Elementos de ingeniería de las reacciones químicas. Pearson Educación. México. pp. 8-11; 34-39.

[51] Jawed, M. and Tare, V (1999) Microbial Composition Assessment of Anaerobic Biomass Through Methanogenic Activity Tests Water S.A 25. 3: 345 – 350.

[52] Lettinga, G., van Velsen, A.F.M., Hobma, S.W., De Zeeuw, W. y Klapwijk, A. (1980) Use of the Upflow Sludge Blanket (UASB) Reactor Concept for Biological Wastewater Treatment, Especially for Anaerobic Treatment. Biotechnol. Bioeng .22: 699-734.

[53] Certucha-Barragán M. T, Duarte-Rodríguez G. R, Acedo-Félix E, Almendariz-Tapia F. J, Monge-Amaya, Valenzuela Garcia J. L, Leal-Cruz A. (2010) Estudio de la Biosorción de Cobre Utilizando Lodo Anaerobio Acidogénico. Rev. Int. Contam. Ambient. 2: 26-36.

[54] Shuler, M. L. and Kargi, F. (2002) *Bioprocess Engineering: Basic Concepts, 2nd edition,*Prentice-Hall. Inc .

[55] Reyes E. D., Cerino C. F. y Suárez H. M. A. (2006) Remoción de metales pesados con carbón activado como soporte de biomasa. Ingenierías. IX, 31.59-64.

[56] Cañizares- Villanueva, R. O. (2000) Biosorción de metales pesados mediante el uso de biomasa microbiana. Revista Latinoamericana de Microbiología. 42:131-143.

[57] Gupta R., Prerna Ahuja, Seema Khan, R. K. Saxena and Mohapatra H. (2000) Microbial biosorbents: Meeting challenges of heavy metal pollution in aqueous solutions. CURRENT SCIENCE, 78, 8, 967-973.

[58] Leyva Ramos R., Medellín C. N. A., Guerrero C. R.M., Berber M. M.S., Aragón P. A. y Jacobo A. A. (2005) Intercambio iónico de plata (I) en solución acuosa sobre clinoptilolita. Rev. Int. Contam. Ambient. 21 (4) 193-200.

[59] Rodríguez, H. (2001). Estudio de la contaminación por metales pesados en la Cuenca de Llobregat. Tesis Doctoral. Barcelona.

[60] Vega B. (2008) Inmovilización de células. 1º Curso Iberoamericano de Biocatálisis Aplicada a Química Verde. Facultad de Química. Universidad Nacional de Quilmes. Argentina.

[61] Krause, U., Thomson-carter, F.M. and Pennington, T.H. (1996) Molecular Epidemiology of *Escherichia coli* O157:H7 by Pulsed-Field Gel Electrophoresis and Comparison with That by Bacteriophage Typing. J.Clin. Microbiol. 34:959-961.

[62] Núñez Gaona O. (2004) Producción de invertasa por *Aspergillus niger* en fermentación en medio sólido. Tesis maestria en biotecnología. UAM-Iztapalapa, México.

[63] Servin R. L. (2006) Metalurgia de minerales no metálicos, zeolitas. http://www.monografias.com/trabajos/zeolitas/zeolitas.htm. Consulta: Abril 2012.

[64] Aguayo, S.S. y Mejía, Z. F. A. (2006) Adsorción de arsénico en zeolitas naturales pretratadas. Memorias del XVI Congreso Internacional de Metalurgia Extractiva. Saltillo, Coahuila. 303-310.

Chapter 7

Short-Rotation Coppice of Willows for the Production of Biomass in Eastern Canada

Werther Guidi, Frédéric E. Pitre and Michel Labrecque

Additional information is available at the end of the chapter

1. Introduction

The production of energy by burning biomass (*i.e.* bioenergy), either directly or through transformation, is one of the most promising alternative sources of sustainable energy. Contrary to fossil fuels, bioenergy does not necessarily result in a net long-term increase in atmospheric greenhouse gases, particularly when production methods take this concern into account. Converting forests, peatlands, or grasslands to production of food-crop based biofuels may release up to 400 times more CO_2 than the annual greenhouse gas (GHG) reductions that these biofuels would provide by displacing fossil fuels. On the other hand, biofuels from biomass grown on degraded and abandoned agricultural lands planted with perennials do not have a negative effect on carbon emissions [1]. In addition, when properly managed, bioenergy can enhance both agricultural and rural development by increasing agricultural productivity, creating new opportunities for revenue and employment, and improving access to modern energy services in rural areas, both in developed and developing countries [2].

Biofuels constitute a very broad category of materials that can be derived from sources including municipal by-products, food crops (*e.g.* maize, sugar cane etc.), agricultural and forestry by-products (straws, stalks, sawdust, etc.) or from specifically-conceived fuel crops. Our analysis focuses on agricultural biofuel crops that can be grown in temperate regions. These crops can be divided into four main categories (Table 1).

Oilseed crops have long been grown in rotation with wheat and barley to produce oil for human, animal or industrial use. Today, these crops primarily provide feedstock for biodiesel. Biodiesel is produced by chemically reacting a vegetable oil with an alcohol such as methanol or ethanol, a process called transesterification. Cereals and starch crops, whose main economical use is for food and fodder, can also be transformed to produce biofuels. For example, the starch in the grains of maize (*Zea mays* L.), wheat (*Triticum aestivum* L.) and

Category	Common name	Botanical name	Habit	Crop life cycle	Main destination
Oil crops	Camelina	*Camelina sativa* (L.) Crantz	Herbaceous	Annual	Biodiesel
	Castor	*Ricinus communis* (L.)		Mostly annual	
	Field mustard	*Sinapis alba* (L.)		Annual	
	Groundnut	*Arachis hypogaea* (L.)			
	Hemp	*Cannabis sativa* (L.)			
	Linseed	*Linum usitatissimum* (L.)			
	Oilseed rape	*Brassica napus* (L.)			
	Safflower	*Carthamus tinctorius* (Mohler)			
	Soybean	*Glycine max* (L.) Merr.			
	Sunflower	*Helianthus annuus* (L.)			
Cereals	Barley	*Hordeum vulgare* (L.)	Herbaceous	Annual	1ˢᵗ gen. ethanol / Solid biofuel
	Maize	*Zea mays* (L.)			
	Oats	*Avena sativa* (L.)			
	Rye	*Secale cereale* (L.)			
	Wheat	*Triticum aestivum* (L.)			
Starch crops	Jerusalem artichoke	*Helianthus tuberosus* (L.)	Herbaceous	Perennial	1st gen. ethanol
	Potato	*Solanum tuberosum* (L.)		Annual	
	Sugar beet	*Beta vulgaris* (L.)		Biennial	
	Sugarcane	*Saccharum officinarum* (L.)		Perennial	
Dedicated bioenergy crops	Kenaf	*Hibiscus cannabinus* (L.)	Herbaceous	Annual	Solid biofuel / 2ⁿᵈ gen. ethanol
	Sorghum	*Sorghum bicolor* (L.) Moench			
	Cardoon	*Cynara cardunculus* (L.)	Herbaceous	Perennial	
	Giant reed	*Arundo donax* (L.)			
	Miscanthus	*Miscanthus* spp.			
	Reed canary grass	*Phalaris arundinacea* (L.)			
	Switchgrass	*Panicum virgatum* (L.)			
	Short-Rotation Coppice	*Eucalyptus* spp. *Populus* spp. *Salix* spp.	Woody	Perennial	

Table 1. The main bioenergy crops for regions with a temperate climate.

sorghum (*Sorghum bicolor* (L.) Moench) can be converted to sugars and then to ethanol by traditional fermentation methods for use in transportation and other fuels (*e.g.* bioethanol). These crops may also be used to produce biogas, composed principally of methane and carbon dioxide produced by anaerobic digestion of biomass. These energy crops have the advantage of being relatively easy to grow. Most are traditional agricultural crops and are easy to introduce at the farm level since they do not require particularly cutting-edge technological equipment. However, using food crops as a source of bioenergy raises serious issues related to food supply and costs, and consequently has been under increasing criticism from the scientific community and society. In particular, the use of these crops for bioenergy competes directly with their use as food. In addition, since many of these crops are annuals, they require large energy inputs and fertilizer for establishment, growth and management, and thus in the end result in minimal energy gains. For such reasons, these crops may not be efficient either for achieving energy balances or for reducing greenhouse gas emissions.

The category of dedicated energy crops notably includes all lignocellulosic (mostly perennial) crops grown specifically for their biomass and used to produce energy. Such crops include herbaceous (*e.g.* miscanthus, switchgrass, reed canary grass, etc.) and woody (willow, poplar, eucalyptus) species that have been selected over the past decades for their high biomass yield, high soil and climate adaptability, and high biomass quality. In addition, especially if grown on marginal arable lands, they do not compete directly for use for food [3], do not require large amounts of inputs in terms of annual cultivation and fertilizer applications [4], nor involve the destruction of native forests with severe negative effects on carbon sequestration [5] and biodiversity [6-7].

We shall limit our description to woody species, because they constitute the focus of our research.

Woody crops for energy production include several silvicultural species notably sharing the following characteristics: fast growth and high biomass yield, potential to be managed as a coppice and high management intensity (highly specific needs with regard to fertilization, irrigation, etc).

A recent review of the literature revealed that about ten different terms are used to refer to the silvicultural practice of cultivating woody crops for energy production: short-rotation woody crops, short-rotation intensive culture, short-rotation forestry, short-rotation coppice, intensive culture of forest crops, intensive plantation culture, biomass and/or bioenergy plantation culture, biofuels feedstock production system, energy forestry, short-rotation fiber production system, mini-rotation forestry, silage sycamore, wood grass [8]. The same author suggested adoption of standard terminology based on an earlier work [9] that had defined this cropping system as *"a silvicultural system based upon short clear-felling cycles, generally between one and 15 years, employing intensive cultural techniques such as fertilization, irrigation and weed control, and utilizing genetically superior planting material"*, to which he proposed to add *"and often relying on coppice regeneration"*, since most species used are able to sprout following harvest. The term coppice refers to a silvicultural practice in which the stem of a tree is cut back at ground level, allowing new shoots to regenerate from the stump.

The early growth rate of coppice sprouts is much greater than that of seedlings or cuttings and in this way trees managed as coppice are characterized by remarkably fast growth and high biomass yield [10-11]. The main species under this cultivation regime in temperate climates are poplar (*Populus* spp) [12], willow (*Salix* spp) [13] and eucalyptus (*Eucalyptus* spp.) [14], and to a lesser extent, black locust (*Robinia pseudoacacia* L.) [15] and alder (*Alnus* spp.) [16]. All of these species, which are cultivated for biomass production in a specific region, are fast-growing under local conditions, cultivated in dense stands (to take maximum advantage of available nutrients and light, resulting in maximum growth), harvested after short rotation periods (usually between 2-8 years), and coppicable (thus reducing establishment costs). In addition, willows and poplars demonstrate ease of vegetative propagation from dormant hardwood cuttings, a broad genetic base and ease of breeding. These characteristics make them ideal for growing in biomass systems and facilitate clonal selection and ensure great environmental adaptability [17].

2. Willow short-rotation coppice in Quebec

2.1. A brief history

Scientific interest in short-rotation bioenergy willows in Canada dates back to the mid-1970s' oil crisis, which stimulated the use of biomass for energy production. The Federal government's 1978 ENFOR (ENergy from the FORest) program, coordinated by the Canadian Forest Service was part of a federal interdepartmental initiative on energy research and development to promote projects in the forest bioenergy sector. Scientists from the Faculty of Forestry at the University of Toronto pioneered the investigation of willow's potential for bioenergy in Canada, convinced that willows could produce high annual yields in temperate zones [18-19] Louis Zsuffa's (1927-2003) work on selection and breeding of poplars and willows through genetic trials on small surfaces inspired the next generation of researchers, including one of his graduate students, Andrew Kenney, who implemented short-rotation intensive culture technology on the first prototype energy plantations in Canada [20]. As well, Gilles Vallée, of the Quebec ministry of Natural Resources, investigated the genetic improvement of hybrid poplar and willow with the aim of developing clones adapted to the shorter growing seasons of boreal forest locations. Our own *Institut de recherche en biologie végétale* (Plant Biology Research Institute), located at the Montreal Botanical Garden, grew out of the ENFOR program in the early 1990'. Our research team initially set out to identify willow species and clones well-adapted to short-rotation coppice in southern Quebec (Eastern Canada). Our experiments showed that Quebec's climate and soil are very favourable for growing various willow clones in short rotation, and that wastewater sludge can be an effective low-cost and environmentally-friendly fertilizer [21]. Researchers from Federal and provincial ministries also initiated diverse willow projects during the 1980s and 1990s, including the genetic improvement of hybrid poplar and willow clones adapted to the short growing seasons of boreal forests [22]. Simultaneously, Natural Resources Canada, a federal ministry, collaborated with several committees, including the International Energy Agency, to improve cooperation and information exchange between countries that have national programs in bioenergy research.

From the early 1990s to the present, dedicated, continuous research on willows in the Canadian context has been concentrated at the Montréal Botanical Garden. As a result of these extensive research efforts, approximately 300 ha of willows have been established on marginal agricultural lands in Quebec over the last 20 years.

2.2. Site selection

Several environmental factors can potentially influence a willow short-rotation coppice plantation and all should be evaluated prior to plantation establishment to maximize success. Ecologically, the majority of willow species are common in cold temperate regions and are adapted to moist-hygric habitats. However, most riparian species require well aerated substrate and flowing moisture, whereas non-riparian species have less exacting soil aeration requirements [23]. Moisture availability is an important factor determining native distribution in natural environments, successful plant establishment and high biomass yield. On average, willow coppice requires more water for growth than conventional agricultural crops [24] and consequently highly moisture retentive soil is an essential prerequisite. The lower St. Lawrence Valley, where most willow plantations in Quebec have been successfully established over the past two decades, is characterized by a temperate and humid climate with an annual average temperature of 6.4°C, average growing season (May-October) temperature of 15.8°C and a mean total annual precipitation of 970 mm. The period without freezing is on average 182 days and the total number of growing degree-days (above 5°C) is 2100.

Soil composition is another important factor for ensuring willow crop establishment and yield. In general, willow can be grown on many types of agricultural land. However, since this species is more water-dependent than other crops, particularly dry land should be avoided. On the other hand, although willow has been shown to be a rather flood-tolerant species compared to other woody energy crops [25], permanently submerged soils also constitute unsuitable sites. Ideally, willows should be grown on a medium textured soil that is aerated but still retains a good supply of moisture. Most willows grow best in loamy soils, with a pH ranging from 5.5 –7.0, although to a certain extent suitable soil types may range from fine sands to more compact clay soils. Several studies have shown that heavy clay soils are not very suitable for willows [26]. Most abandoned agricultural lands in Quebec are thus highly suited to growing willows, being situated in temperate regions and often adequately fertile. Other pre-establishment considerations are linked to the location of the plantation. Economical (and ecological) benefits can be maximized when high production levels of willows are achieved in combination with low input requirements, which result in high-energy efficiency and low environmental impact. For this reason, choosing the right location is crucial for achieving a sustainable energy production system. Normally, the plantation should be situated as close as possible to the end utilisation point (*e.g.* within 50-100 km from a power plant or transformation industry, etc.) and in any case should be established in proximity to main roads, highways or railroads. For the same reasons, the shape of willow fields should be as regular as possible to avoid loss of time and energy during management and harvest operations. For practical reasons (mainly linked to tillage and harvest) land with an elevated slope (>15%) should be avoided. Ideal sites are flat or with a slope not exceeding 7-8%.

2.3. Choice of planting material

Willow yield varies greatly depending on both environmental and genetic factors. The genus *Salix*, to which willows belong, comprises 330 to 500 species worldwide of deciduous or, rarely, semi-evergreen trees and shrubs [27] and the number and variety of species along with the ease of breeding have facilitated clonal selection adapted to several goals (ornamental, silvicultural, environmental applications, etc.). However, a large number of willow species are not suitable for biomass production because of their slower growth rate. Nowadays, the exploitation of the wide biological diversity within the genus *Salix* is focused primarily on a few species (*S. viminalis*, *S. purpurea*, *S. triandra*, *S. dasyclados*, *S. eriocephala*, *S. miyabeana*, *S. purpurea*, *S. schwerinii*, and *S. sachalinensis*), whereas there has been a recent increase in the number of selected intra- and interspecific hybrid cultivars offering higher yields, improved disease resistance and tolerance of a higher planting density (Table 2).

In Quebec, the first trials for evaluating willow biomass potential began on small plots in the early 1990s with two species, one indigenous (*S. discolor*) and the other a European cultivar (*S. viminalis* 5027). Two growing seasons after establishment, their total aboveground biomass yield was very similar – between 15 and 20 t ha^{-1} of dry-matter per year, confirming the high potential of these two species under Quebec's agro-ecological conditions [28]. A subsequent trial aimed at evaluating these two species comparatively with *S. petiolaris* Smith; both the first-tested species were shown superior to the latter in terms of biomass productivity [21]. However, since after a number of years this *S. viminalis* cultivar showed sensitivity to insect attacks, particularly to the potato leaf hopper, and since the risk of epidemic diseases increases as the plantation area expands, a new set of selected clones was investigated. These experiments showed that in contrast to *S. viminalis'* poor performance due to high sensitivity to pests and diseases, other willow cultivars (*S. miyabeana* SX64 and *S. sachalinensis* SX61) could achieve high biomass yields [29]. Now, 10 years later, *S. miyabeana* (SX64) and *S. sachalinensis* (SX61) cultivars still provide the highest biomass yield and greatest growth in diameter and height among willows in the Upper St. Lawrence region. However, selected cultivars from indigenous (*i.e.* North-American) willow species, especially *S. eriocephala* (cultivars S25 and S546) and *S. discolor* (cultivar S 365), perform well and only slightly below SX64, thus making them preferable for use on large-scale plantations in Quebec due to their less rigorous maintenance requirements and sensitivity to insect and pest attacks.

New selected planting material has also been made extensively available by several willow growers interested in development of willow cultivation in Quebec and operating jointly with researchers. Agro Énergie (www.agroenergie.ca) was the first large-scale commercial nursery in Quebec to produce diverse varieties of willow and has continued to expand its willow plantations across Eastern Canada. For the joint project between our research team and Agro Énergie, we provide scientific expertise in terms of plantation layout, species selection, cultivation methods and management practices. The 100 hectares of land provided by Agro Énergie represent an opportunity to scale up experimental technology, perfect techniques and evaluate costs and yield, using the high performance agricultural equipment necessary for large-scale commercial production.

Taxon	English common name	Origin	Comercial varieties and hybrids
S. nigra Marshall	Black willow	North America	S05*
S. triandra L	Almond-leaved willow	Eurasia	Noir de Villaines+, P60101,
S. alba L.	White willow	Europe, Africa, & west Asia	S44*
S. eriocephala Michx.	Heart-leaved willow	North America	S25*, S546*
S. discolor Muhl.	American pussy willow	North America	S365ᴴ¥
S. dasyclados Wimm.	Wooly-stemmed willow	Eurasia	SV1*¥
S. schwerinii Wolf	Schwerin willow	East Asia	
S. udensis (sin S. sachalinensis)Trautv.		East Asia	SX61*
S. viminalis L.	Common osier or basket willow	Eurasia	SVQ*, S33*,5027*, Jorr+
S. miyabeana Seemen	Miyabe willow	East Asia	SX64*, SX67*
S. purpurea L.	Purple willow or purple osier	Northern Africa & Europe	Fish Creek*
S. acutifolia Willd.	Pointed-leaf willow	Eastern Europe	S54*
S. sachalinensis x *S. miyabeana*			Sherburne*, Canastota*
S. purpurea x *S. miyabeana*			Millbrook*
S. eriocephala x *S. interior* *S. viminalis* x *S. schwerinii*			S625* Bjorn+, Tora+, Torhild+, Sven+, Olof+

Table 2. Most common *Salix* taxa and corresponding commercial varieties for biofuel production in Quebec (* Selected in North America; + Selected in Europe;¥ Its identity is currently under study).

2.4. Land preparation and weed control

Appropriate soil preparation is essential to ensure good plant establishment and vigorous growth. This is particularly true when willows are to be established on soil with low fertility or marginal land. The main goal of any land preparation operation should be to eliminate weeds, aerate soil and create a uniform soil surface for planting. Once the planting site has been chosen, the first operation to be performed is preparation of the land much as for any other agricultural crop. The productivity of trees under short-rotation intensive culture is strongly influenced by herbaceous competition. One of the first trials conducted by our research team in the early 1990s showed that weed suppression was essential to willow establishment [30]. On Quebec's generally well-drained lands, the most common weeds are broad-leaved annuals such as white goosefoot (*Chenopudium album* L.) and redroot pig-weed (*Amaranthus retroflexus* L.), whereas on poorly drained lands, annual grasses, barnyard grass (*Echinochloa crusgalli* L.) and perennials such as Canada thistle (*Cirsium arvense* L.) and quack grass (*Agropyron repens* (L.) Beauv.) are more common [30]. In the case of abandoned agricultural lands or in the presence of a high concentration of weeds, one or two applications of a systemic herbicide (*e.g.* glyphosate 2- 4 L/ha) during the summer of the year prior to planting are strongly recommended to promote establishment. A few weeks later, the destroyed plant mass should be incorporated into the soil using a rotating plough. In Quebec, a first ploughing should be performed in the fall prior to planting. Autumn ploughing allows the soil to break down over the winter, and also increases the amount of moisture in the planting bed. Suitable equipment includes any common mouldboard, chisel or disc plough (20 – 30cm depth), following usual agronomical practices for other crops (*e.g.* maize). Power harrowing (15- 18 cm depth) or cross disking of the site should be carried out in the spring immediately prior to planting to ensure a flat, regular planting bed.

2.5. Plantation design and planting

Willows can be planted according to two different layouts. In most North European countries (Sweden, UK, Denmark) and in the US, the most frequent planting scheme is the double row design with 0.75 m distance between the double rows and 1.5 m to the next double row, and a distance between plants ranging from 1 m to 0.4 m, corresponding to an initial planting density of 10,000 - 25,000 plants ha⁻¹. The most common plantation density in these countries is currently around 15,000 (1.5 x 0.75 x 0.59 m) plants ha⁻¹ [31]. This rectangular planting arrangement is used to facilitate field machine manoeuvres through the plantation site. Tractors overlap the double row and the wheels run in the wider strips between those rows [32]. In Quebec, a simpler willow planting design, similar to that used for poplar in short rotations, has been in use since initial trials with only minimal modifications. It consists of a single row design ranging from 0.33 m between plants on a row and 1.5 m between rows (20,000 plants ha⁻¹) in the very first plantations, to 0.30 m on the row and 1.80 m between rows (18,000 plants ha⁻¹) in newer willow plantations. Theoretically, this design facilitates weed control during the establishment phase (the first three years), and consequently willow rooting and growth. In fact, the design choice depends mostly on machinery available for planting and harvesting, since it has been clearly demonstrated that planting design has less impact than plant density and cutting cycle on the yield of *Salix* plants, due to their ability to take advantage of the space available to each stool [32]. The choice of planting density must take into account other ecological factors as well. On sites with appropriate water supply, plantation establishment and subsequent biomass production depend largely on agronomic considerations such as plant spacing and harvesting cycles. Many studies have reported a correlation between spacing and harvesting cycles. In general, maximum yields are achieved early in dense willow plantations, but wider spaced plantations ensure the highest long-term biomass yield [33-34]. On the other hand, under short harvesting cycles, willow stands have a shorter duration, as they are likely to be more exposed to pathogens [35]. At present, most willow short-rotation stands in Quebec have a plantation density of about 16,000 to 17,000 cuttings ha⁻¹ and are harvested every two to three years.

Planting material consists of dormant willow stem sections, either rods or cuttings, depending on the planting machinery to be adopted. In some countries, for example in the UK and in the USA, 'step planters' are the most commonly used machines. Willow rods 1.5-2.5 m long are fed into the planter by two or more operators, depending on the number of rows being planted. The machine cuts the rods into 18-20 cm lengths, inserts these cuttings vertically into the soil and firms the soil around each cutting. Step planters have been calculated to cover 0.6 ha/hr in a UK study. [31]. In Quebec, the most common planting machine is a cutting planter that uses woody cuttings (20-25 cm long) and may operate on 3 rows simultaneously (Figure 1).

Figure 1. Willow planting machine operating on 3 rows simultaneously

Normally, a cutting planter inserts cuttings into the soil at a depth of about 18 cm. Based on empirical experience, this equipment can plant 3,600-4,000 cuttings per hour (1 ha of willow every 3-4 hours), although the duration of this operation may vary depending on several factors (site topography, soil type, plot shape, etc.). Planting material in Quebec is prepared by harvesting one-year-old stems (about 3 m long) in the autumn (*i.e.* when plants are dormant) of the year prior to planting. This material is wrapped in plastic film to avoid moisture loss, and stored in a refrigerator at -2 to -4°C. In spring, two to three weeks prior to planting, healthy willow rods 1-2 cm in diameter (with no symptoms of disease on bark or wood) are selected to prepare cuttings. Tips of stems bearing flower buds are first discarded. Then the rest of the whip is cut into 20-25 cm lengths using an adapted rotary saw and stored in boxes, ready to be planted (Figure 2).

Figure 2. Willow cuttings before planting

If cuttings are left in temperatures above 0°C, a break in their dormancy will occur, adventitious roots will develop and the buds may burst. This will lead to a reduction in water and nutrient content and consequently reduced viability. Thus, it is very important to plan the planting operation carefully in advance, calculating the number of cuttings that can be planted.

The time of planting varies according to meteorological and soil conditions. Planting should be undertaken as soon as possible in the spring, to allow plants to benefit from the high soil water content following snowmelt, and then to establish quickly and take maximum advantage of a long growing season. In addition, a late willow planting is also more subject to failure due to drought if a dry summer should occur. However, there are several additional factors that play an important role in determining the planting date. In order for soil preparation (*e.g.* harrowing) to begin in the spring, soil should be free from snow but not so muddy that soil structure could easily be damaged by tractors. The date at which such conditions are met vary considerably from year to year, but in southern Quebec, it usually falls during May, although late planting (up to mid-June) is possible and, in our experience, does not result in serious problems in plant establishment. Planting willow in the colder, northernmost regions of Quebec (*e.g.* Abitibi) may take place up to the beginning of July. In all of these situations, rapid colonisation by highly competitive weed species occurs on fertile sites, thus the use of appropriate residual herbicides is essential to maximize plant survival and early growth. Pre-emergence residual herbicide should be applied immediately upon completion of planting (within a maximum delay of 3-5 days). A mixture of two herbicides (2.30 kg Devrinol and 0.37 kg Simazine per hectare) has been effective on most of our plantations. Since the treatment must reach the zone of weed seed germination, most pre-emergence herbicides require mechanical incorporation (such as by a power tiller) as well as adequate irrigation or natural moisture (rainfall or snow) for best results. More recently, a new herbicde (SureGuard, a.i. flumioxazin) has received approval for pre-emergent use at the time of planting on poplar and willow (including planting stock production in the field, on both stoolbeds and bareroot beds).

2.6. Crop management

2.6.1. Establishment year

All operations carried out in a willow stand during the first year are aimed at promoting plant establishment and a high survival rate, thereby ensuring the on going productive life of the plantation. Weeds are the main problem encountered in willow crop, and they may still colonise fields despite pre-emergence treatments. It was established decades ago that during the first year after planting, vigorous weeds reduce willow growth by between 50% and 90% [36]. Most of these invasive species have higher growth rates than young willow shoots, and compete with them mainly for light [37], and to a lesser extent for water and nutriments, leading to high plant mortality within the first few months. Hence, great care should be taken to control weed development in the field in the weeks following planting. On most willow plantations in Quebec, one to three passes with a rotary tiller cultivator between rows are needed to control weeds during the establishment year. In case of a severe weed problem, manual weeding may be required between plants within each row.

2.6.2. Cutback

There is much evidence that most newly-established willow plantations profit immensely from being cut back at the end of the first growing season (Figure 3).

Figure 3. After cutback willows sprout vigorously from the stumps

Not only does cutback encourage established cuttings to produce vigorous multiple shoots the following spring, it also helps reduce competition by weeds, thereby reducing the need for continued chemical weed control [38]. Furthermore, cutback facilitates entering the field at the beginning of the second growing season to fertilize and till soil between rows. Cutback is normally performed in the fall by cutting all newly-formed shoots at ground level using conventional agricultural equipment, such as reciprocating mowers for large surfaces or a trimmer/brush-cutter for small plots.

2.6.3. Fertilization

For many reasons, fertilization is a controversial aspect of short-rotation plantation, subject to fluctuations in practice. Our review of the historical evolution of willow short-rotation forestry in different countries suggests that the initially highly favourable attitude toward using chemical fertilizers has tended to attenuate over time, mainly because other issues beyond the biomass yield (both economical and environmental) have arisen. Different perspectives on this topic have also arisen out of legislation that in some countries has favored more environmental-friendly management (e.g. by reducing mineral fertilization and enhancing the application of biosolids and waste materials) of bioenergy cropping systems.

However, it is an irremediable fact that, due to high biomass yields, most willow energy crops grown in short-rotation and intensively managed and harvested remove nutrients at a high rate, though evidence varies somewhat (Table 3).

Annual nutrient removal (kg tDM^{-1})					
N	P	K	Ca	Mg	Reference
20.6	6.9	13.7	-	-	[39]
13.6	1.5	8.5	-	-	[40]
13.0	1.6	8.3	-	-	
6.3	1.0	7.5	-	-	[41]
5.7	1.0	3.0	3.0	1.0	[42]
5.3	0.9	3	7.2	0.7	
7.5	0.6	1.8	4.2	0.4	[43]
5.0	0.7	1.8	3.5	0.3	
3.9	0.5	1.5	3.6	0.2	[44]
3.5	0.5	2.5	-	-	[45]

Table 3. Average mass of nutrient removal (kg) per oven dry ton of aboveground willow biomass

Some authors have highlighted that N fertilization in willow plantations at the beginning of the cutting-cycle, excluding the year of planting, is generally a very efficient way to enhance plant growth [45-46]. On the other hand, willow nutrient requirements are relatively low, due to efficient recycling of N from litter and the relatively low nutrient content retained in biomass (stem). Therefore, much less nitrogen fertilizer should be applied than is typical with agricultural crops, although dosage should also be based on formal soil chemical analyses performed prior to plant establishment. Several authors have indicated that no nitrogen is required in the planting year for short-rotation coppice [39-47]. This also reduces the competitiveness of weeds that would take advantage of fertilizer application. Economical considerations are yet another factor to consider when determining the dose of fertilizer to be used, since fertilizer constitutes a significant percentage of the financial cost involved in the production of willow biomass crops. A recent study conducted in New York

State showed that fertilizer represents up to 10–20% of the cost of production over several rotations [48]. The average dose generally recommended in Quebec ranges from the equivalent of 100-150 kg N, 15 kg – 40 kg P and around 40 kg K per hectare per year after the establishment year. Because it is not possible to introduce heavy equipment into the field after plantation establishment, fertilizer application is normally performed one year after planting and after any harvest, when tractors can circulate freely in the field.

An interesting alternative to mineral fertilizers are biosolids and other industrial and agricultural byproducts, which have been tested in many countries since the early 1990s. These include municipal wastewater [49], wastewater from the dairy industry, landfill leachate [50], diverted human urine [51], industrial wastewaters such as log-yard runoff [52], as well as solid wastes like digested or granulated sludge [53] and pig slurry [54]. In fact, the majority of these products contain high levels of nitrogen and phosphorous, elements that might constitute a source of pollution for the environment but at the same time represent a source of nutrients for the plant. Thus there are many advantages to using such products in willow plantations:

1. recycling of nutrients, thereby reducing the need for farmers to invest in chemical fertilizer;
2. conservation of water;
3. prevention of river pollution, canals and other surface water, into which wastewater and sewage sludge would otherwise be discharged;
4. low-cost, hygienic disposal of municipal wastewater and sludge.

Willow cultivated in short rotation is a very suitable crop for fertilization with these products for several reasons. First, it has been determined, both by measured and estimated models, that this crop has high evapotranspiration rates and thereby consumes water quantities as high as any other vegetation cover, which allows significant wastewater disposal over each growing season [24-55-56]. Furthermore, willow short-rotation stands have been shown to be able to uptake large amounts of nutrients present in this waste [57]. Last but not least, willow coppice is a no food no fodder crop and, if properly handled, any possible source of human or environmental contamination is strongly reduced [58]. In some early trials carried out in Quebec to test the possibility of using sludge in willow short-rotation culture, it was found that a moderate dose of dried and palletized sludge (100-150 kg of "available" N ha^{-1}) might constitute a good fertilizer during the establishment of willows, especially on clay sites [53-59]. Today, the recommended dose of derived wastewater sludge fertilizer in Quebec ranges between 18-21 t ha^{-1}of dried material, which corresponds to 100-120 kg available nitrogen per hectare. Fertilization is performed in spring of the second year after planting with ordinary manure spreading machines. Another recent project investigated the effect of the use of pig slurry as fertilizer on the productivity of willow in short-rotation coppice (Figure 4).

The results showed that pig slurry is good fertilizer for willow plantations [54]. In fact, very high biomass yields were obtained over two years, and even made it possible to predict that typical three-year rotation cycles could be reduced to two years, under the proper

Figure 4. Pig slurry application to a willow plantation

production conditions. This means that even though nitrogen in slurry may be less efficient than that in a mineral fertilizer, a significant reduction in the production costs of willow-based biomass as well as recycling of a greater quantity of slurry can be achieved simultaneously [54].

2.7. Pests and diseases

Although there are a great number of insects feeding on willows, three main species are of concerns for willow short rotation coppice in Quebec. The first is the willow leaf beetle (*Plagiodera versicolora* Laicharteg.), one of the most common insects found on willows. The willow leaf beetle is a small (4 - 6 mm long), metallic-blue beetle widely distributed around the world. In Quebec, adults emerge from their overwintering quarters under the loose bark and feed on young willow foliage in spring. Egg laying begins in mid-June. Females lay yellow eggs grouped on the undersides of the leaves. The young larvae emerge a few days later and begin feeding on both sides of the leaves and eating the tissue between the veins, thus skeletonizing the leaves and, depending on the extent of the attack, in all probability leading to a reduction of plant growth. In Quebec, this insect has been frequently observed feeding on leaves of clones of *Salix viminalis* and to a much lesser extent on most common commercial varieties of *S. miyabeana* (SX64 and SX67) and *S. sachalinensis* (SX61). To date, the reported threshold of damage caused by this insect has never been high enough to justify any type of control. However, in case of severe attack, non-toxic products based on *Bacillus thuringiensis*, shown to be effective in eliminating this pathogen, can be used [60].

The other predominant insects found feeding on willow trees and shrubs are two aphid species: the giant willow aphid, *Tuberolachnus salignus* (Gmelin) and the black willow aphid, *Pterocomma salicis* (L) [61].

The giant willow aphid is one of the largest aphids ever recorded, measuring up to 5.8 mm in length [62]. It feeds almost exclusively on willow, but has very occasionally been recorded on poplar (*Populus* spp.). The species is strongly aggregative, forming vast colonies on infested trees. These colonies can cover a significant portion of the 1-3 year old stem surface of a willow tree. Laboratory experiments with willows grown in soil and in hydroponic culture have revealed that this species can reduce the above-ground yield of biomass willows, have severe negative effects on the roots and reduce the survival of both newly planted and established trees [63]. Other preliminary studies carried out in the UK have shown that this insect's feeding behavior is affected by chemical cues from the host. Researchers found that one of its most preferred willows was *S. viminalis* [64]. Although large colonies of this insect have recently been found on several willow varieties in Quebec, it is not yet possible to estimate its threat to willow plantations in this region (Figure 5).

Figure 5. Giant aphids feeding on willow. This insect is often found forming large colonies at base of the stem.

The black willow aphid, *Pterocomma salicis* (L) may actually pose a threat only if severe, frequent attacks occur. Several studies have shown that this species is less damaging than the giant willow aphid, with a less persistent negative impact on willow growth. In Quebec, high density populations of this species have recently been found at the end of June on a willow plantation in the upper St. Lawrence River valley (Huntingdon), mainly on *S. miyabeana* (SX67 and SX64); it did not seem to feed on *S. viminalis*.

Other less damaging insects have been found on willow plantations in Quebec. *Calligrapha multipunctata bigsbyana* adults and larvae may feed on willow leaves without destroying leaf veins, with consequences quite similar to those of *Plagiodera versicolora*. Willow flea beetles of the genus *Crepidodera* (*C. nana* and *C. decora* also feed on *Salicaceae* leaves [65], and are easy to recognize by their brilliant metallic and bicoloured upper surface; blue or green head and pronotum tinged with strong bronze, copper or violet; and unicolorous blue or green elytra. This beetle feeds on either the upper or lower leaf surface, consuming the epidermis and tissue below, but not on the opposite side. After desiccating, the tissue falls out, resulting in a leaf with a bullet-hole appearance. Varieties of willows developed in Europe, based on pedigrees with *Salix viminalis* or *S. viminalis* x *S. schwerinii*, are susceptible to potato leafhopper (*Empoasca fabae* Harris), which causes serious damage to this species and its cultivars or hybrids. Willow shoot sawfly (*Janus abbreviates* Say) larvae have recently been found in Quebec, carving deep tunnels on young willow *S. miyabeana* SX64 shoots where they cause wilting, change of colour (brown or black) and eventually drooping of shoot tips. It has been observed that in some cases 30% of individuals of SX64 in Huntingdon showed at least one shoot affected by this insect. However, only repeated and severe attacks in young willow plantations may adversely affect tree growth.

Willow can be injured by several diseases [66]. Willow leaves may be sensitive to *Alternaria* spp., *Melamsora* spp. and *Venturia* spp., whereas *Cryptodiaporthe* spp., *Glomerella* spp. and *Valsa* spp. are found to affect stems and branches and *Armillaria* spp., *Fusarium* spp. and *Verticilium* spp. roots [67]. However, the most widespread, frequent and damaging disease in willow plantations is leaf rust, caused by *Melampsora* spp. In northern Europe, leaf rust is considered a major factor limiting growth of short-rotation coppice willow [68]. It can cause premature defoliation, poor cold acclimation, premature leaf senescence, and a predisposition to abiotic stress (*e.g.,* competition and drought) in host trees, along with secondary disease organisms, and it may reduce yields by as much as 40% [69]. One of the main alternative solutions to spraying fungicides proposed in northern Europe is growing willow in inter- and intra-species mixtures [70]. If a variety dies out of a mixture due to disease, competition or some other factor, the remaining varieties can compensate for the loss [71]. In some willow plantations in Quebec, severe attacks of *Melamsora* spp. have been detected mainly on a specific commercial clone S301 (*S. interior* 62 x *S. eriocepala* 276), which seemed to be more vulnerable to rust than any other clone studied in the area [29]. Few rust attacks have been reported for most commercial clones, however, chemical or biological disease control is generally not required.

2.8. Harvesting and yields

Willow should be harvested at the end of each rotation cycle (2-5 years), normally in fall, after leaf shedding. All willow stems should be cut at a height of 5 - 10 cm above the soil surface in order to leave a stump from which new buds will form sprouts the following spring. Essentially, there are three ways to harvest willows, the choice largely depending on the final destination of biomass and the equipment available. When willows are grown to produce rods to be used in environmental engineering structures such as sound barriers, snow fences and wind breaks along highways and streets [72-73] or to produce new cuttings, plants are harvested with trimmer brush-cutters. Whole willow rods can also be stored in heaps at the edge of the field and chipped after drying.

Another option involves the use of direct-chip harvesting machines (*e.g.* Class Jaguar and Austoft). This technique uses modified forage harvesters specifically designed to harvest and direct chip willow stems: the stems are cut, chipped and dropped into a trailer either driven parallel to the harvester or connected directly to it. Although this harvest model is very economically efficient and recommended in many countries, it also presents several disadvantages that should be carefully evaluated. Willow biomass has a moisture content of 50-55% (wet basis) at harvest. Consequently, storage and drying of the freshly chipped wood may cause problems. It has been shown that stored, fresh wood chip in piles can heat up to 60°C within 24 hours and start to decompose. Biomass piles require careful management because internal fermentation can cause combustion and the high level of fungi spore production can lead to health problems for operators. Decomposition processes cause a loss of biomass of up to 20% and a significant reduction in calorific value (*i.e.* energy value) of the biomass [74]. Thus, this type of harvest system requires infrastructures to mechanically dry the biomass (*e.g.* ventilation, heating, mixing machinery) and these post-harvest operations increase the production cost. Alternatively, the freshly chipped material should be delivered to heating plants as soon as possible.

A third harvest system recently developed in Canada, mainly adapted to willow short-rotation coppice, is a cutter-shredder-baler machine that performs light shredding and bales willow stems [22], producing up to 40 bales ha^{-1} (20 t ha^{-1}) on willow plantations (Figure 6).

Figure 6. Willow cutter-shredder-baler harvester operating in Quebec

The main advantage is that, since bales can be left to dry before being chipped, the risks linked to handling wet biomass are reduced [75]. In Quebec, willow biomass harvest is usually done in fall after leaf shedding.

As with any other agricultural crop, biomass yield of willow short-rotation coppice depends on many co-occurring factors including cultivar, site, climate and management operations. Soil type, water availability, and pest and weed control also affect yield. Data from existing commercial sites in the UK suggest that average yields of around 8-10 odt ha⁻¹yr⁻¹ are representative of plantations using older cultivars, whereas biomass yields as high as 15-18 odt ha⁻¹yr⁻¹ can be obtained by using selected genetic material [31]. In other northern European countries, an average annual growth of 15–20 odt ha⁻¹yr⁻¹ has been observed in early experiments [76], although more recent figures suggest that an average of 10 odt ha⁻¹yr⁻¹ is more realistic [77]. Experimental yields of short-rotation willow ranging from 24 to 30 oven dry tonnes (odt) ha⁻¹ yr⁻¹ have been measured in the US and Canada [43-44], although typical yields are more often in the range of 10 to 12 odt ha⁻¹ yr⁻¹ [78].

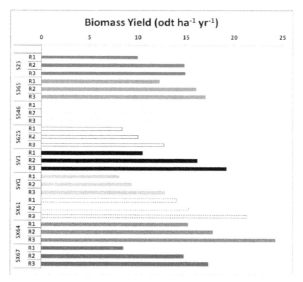

Figure 7. Average biomass yield for nine willow cultivars during three successive rotations (10 years) in the Upper St. Lawrence region (Quebec) on former farmlands. Clones SX64 and SX61 along with some indigenous species (S25, S365, S546) are the most productive and thus are considered to be very suitable for short-rotation forestry in southern Quebec.

Long-term trials show that under southern Quebec's pedoclimatic conditions, short-rotation willow coppice can provide high biomass yields over many years, although results vary according to variety. In one clonal test for instance, at the end of the third (3-years) rotation cycle, the most productive willow cultivars were SX64 (19 Odt ha⁻¹ yr⁻¹) and SX61 (17 Odt ha⁻¹ yr⁻¹) (Figure 7). Also, indigenous (*i.e.* North-American) willow cultivars, especially *S. eriocephala* (S25 and S546) and *S. discolor* (S 365) cultivars, show high biomass potential (13 - 15 Odt ha⁻¹ yr⁻¹). A scientific follow up of an old willow plantation established in Huntingdon in southern Quebec (Canada), showed that willows were still able to maintain a high level of productivity after five coppicing cycles. Plants can remain vigorous and produce high yields (14 Odt ha⁻¹ yr⁻¹) even after 18 years of cultivation (Table 4). This represents a very important demonstration of the viability of long-term economic exploitation of willows.

	Average biomass yield	
Rotation	Total (Odt ha⁻¹)	Annual (Odt ha⁻¹ yr⁻¹)
First (1195-1997)	45.3	15.1
Second (1998-2001)	88.1	22
Third (2002-2004)	51.7	17.2
Fourth (2005-2008)	67.4	16.9
Fifth (2009-2011)	42	14

Table 4. Average biomass yield for *Salix viminalis* L. (clone 5027) achieved during five successive rotations in southern Quebec (Canada)

3. Perspectives for future research: The use of willows in phytoremediation

In Canada, it is estimated that millions of hectares of arable land lie uncultivated. These so-called marginal lands tend to be less productive, less accessible, poorly drained, or even contaminated [79]. Willows have been successfully used to capture leached nutrient and heavy metals from soils [54, 59, 80, 81]. The various species of *Salix* have been shown to establish well on these marginal and contaminated soils, which provides new research opportunities for future applications.

3.1. Phytoremediation

The main types of contaminants found in Quebec soils are petroleum products and heavy metals [82]. In many urban areas, past industrial activities have resulted in thousands of contaminated sites that require decontamination prior to any further utilization. Estimates by the province's ministry of environment have shown that, in the region of Montreal alone, there are over 1350 contaminated sites of which only 54% are in the process of being rehabilitated by traditional methods [83]. Current decontamination methods imply the excavation of the contaminated soils, transport to a landfill treatment facility followed by chemical cleaning, vitrification, incineration or dumping; these steps are extremely expensive [84]. Plant-based *in situ* decontamination technologies, *i.e.* phytoremediation, represent a cost-effective alternative [84]. Plants have the capacity to accumulate, translocate, concentrate, or degrade contaminants in their tissues. Phytoremediation takes advantage of the microbial communities (bacteria and fungi) present in soils to increase the potential of plants to uptake pollutants from the soil matrix. Willows are among the species most widely used for phytoremediation, given their diversity and tolerance of high levels of contaminants [85]. Also, willows develop an extensive root system that stimulates rich and diverse microbial communities that are involved in the degradation of organic pollutants, These characteristics, combined with exceptionally high biomass production, make them very suitable for phytoremediation [86].

Phytoremediation using willows is becoming an increasingly popular alternative approach to decontamination, and several studies and pilot projects are underway. Willows have been used successfully to treat highly toxic organic contaminants such as PCBs, PAHs, and nitro-aromatic explosives [87]. Similarly, willows, in particular *S. viminalis* and *S. miyabeana*, have been shown to accumulate Cd and Zn in their stems and leaves while sequestering Cu, Cr, Ni and Pb in their roots [85,88,89,90]. In previous studies, the efficiency of willows in short-rotation intensive plantation for the elimination of heavy metals contained in wastewater sludge has been investigated [28, 59, 90]. We have also found that willow may be useful for improving sites polluted by mixed organic-inorganic pollution [91] (Figure 8).

Figure 8. Phytoremediation using willows on a former oil refinery around Montreal

Although the fast-growing perennial habits of short-rotation coppice willow planted at high densities result in a low concentration of metals accumulated in biomass after one year of growth, the high biomass production of *Salix* spp. over several harvesting cycles (2-3 years) allows them to accumulate large quantities of metals over the long-term, suggesting great potential as a phytoremediation tool.

3.2. Genetic improvement of willow for phytoremediation

Historically, most genetic selection to improve willow germplasm has been oriented toward increased capacity for biomass production [92], adapted to temperate climates and resistant to pathogens. However, in the context of phytoremediation, the ideal willow genotype must also: i) be adapted to specific pedo-climatic conditions; ii) be fast growing; ii) produce a large root biomass; iv) be resistant to a variety of contaminants; v) have a high concentration factor of contaminants; vi) be easy to establish, maintain and collect. The exceptional diversity of the genus *Salix* makes it an ideal candidate for breeding programs seeking to develop cultivars more efficient at phytoremediation.

To our knowledge, one of the rare efforts to understand the genetic and genomic bases underlying the potential of willow for phytoremediation is the three year Genorem project (www.genorem.ca) launched by research teams at the Université de Montréal and McGill University (Project Leaders Dr. B. Franz Lang and Dr. Mohamed Hijri, both of the Université de Montréal) and involving over thirty scientists, students and staff. The project integrates traditional field and molecular biology experiments, employing recently developed life science technologies: genomics, proteomics, metabolomics and bioinformatics. GenoRem's objectives include the development of guidelines for phytoremediation procedures respectful of the environment that will ultimately be useful to both government and corporate sectors. The transcriptomes of 11 willow genotypes will be sequenced, resulting in basic molecular information about the genes activated in willow when in presence of soil contaminants. GenoRem will also investigate the close relationship established between the willow cultivars studied and the associated soil microorganisms. Ultimately, project results will provide willow breeders with gene markers linked with increased phytoremediation potential.

Phytoremediation as a decontamination technology can be applied to large surface areas, causes less environmental disturbances and represents a significantly cheaper approach than traditional methods. However, treatment is lengthy (several years), and the methodologies appropriate for each type of contamination require refinement. While the biomass produced in the context of a phytoremediation project may potentially be contaminated, this does not affect its utilization as a product outside the food chain. Moreover, the highly concentrated ashes resulting from conversion of the biomass to fuel facilitate disposal and treatment of the contaminant, particularly for a large, diluted volume of contaminated soil. Hence the decontamination by means of phytoremediation is a less intensive technique.

4. Conclusions

Eastern Canada is one region where willow short-rotation coppice has been the focus of numerous research projects over the last 15-20 years. Most experimental data published during this period concerning Quebec have found a high biomass potential, due to a combination of several factors, including the very high biomass yield of certain willow varieties, favourable pedoclimatic conditions and the very low incidence of severe pests and diseases. These high biomass yields have encouraged some growers to choose willows as an alternative agricultural crop, leading to a dramatic expansion of land devoted to willow short-rotation coppice in the province, especially over the last five years. However, the future evolution of this crop's production will most certainly be influenced by the development of an active market for such biomass, which would encourage farmers to grow willow over a much larger surface area. In particular, developments in the technology of feedstock transformation and marketing issues related to product potential both merit further study. The high potential of willow for bioenergy production and environmental applications, including phytoremediation, in the Quebec context has been clearly demonstrated.

Author details

Werther Guidi, Frédéric E. Pitre and Michel Labrecque
Institut de Recherche en Biologie Végétale (IRBV – Plant Biology Research Institute) – Université de Montréal – The Montreal Botanical Garden, Montréal, Canada

5. References

[1] Fargione, J., Hill, J., Tilman, D., Polasky, S., Hawthorne, P., 2008. Land Clearing and the Biofuel Carbon Debt. Science 319, 1235-1238

[2] FAO, 2012. Impacts of Bioenergy on Food Security – Guidance for Assessment and Response at National and Project Levels., Environment and Natural Resources Working Paper, Rome, p. 64.

[3] Tilman, D., Socolow, R., Foley, J.A., Hill, J., Larson, E., Lynd, L., Pacala, S., Reilly, J., Searchinger, T., Somerville, C., Williams, R., 2009. Beneficial Biofuels—The Food, Energy, and Environment Trilemma. Science 325, 270-271.

[4] Nassi o Di Nasso, N., Guidi, W., Ragaglini, G., Tozzini, C., Bonari, E., 2010. Biomass production and energy balance of a twelve-year-old short-rotation coppice poplar stand under different cutting cycles. Global Change Biology Bioenergy 2, 89–97.

[5] Sartori, F., Lal, R., Ebinger, M.H., Parrish, D.J., 2006. Potential Soil Carbon Sequestration and CO2 Offset by Dedicated Energy Crops in the USA. Critical Reviews in Plant Sciences 25, 441-472.

[6] Londo, M., Dekker, J., Ter Kreus, W., 2005. Willow short-rotation coppice for energy and breeding birds: An exploration of potentials in relation to management. Biomass and Bioenergy 28, 281-293.

[7] Bellamy, P.E., Croxton, P.J., Heard, M.S., Hinsley, S.A., Hulmes, L., Hulmes, S., Nuttall, P., Pywell, R.F., Rothery, P., 2009. The impact of growing miscanthus for biomass on farmland bird populations. Biomass and Bioenergy 33, 191-199.

[8] Dickmann, D., 2006. Silviculture and biology of short-rotation woody crops in temperate regions: Then and now. Biomass and Bioenergy 30, 696-705.

[9] Drew, A.P., Zsuffa, L., Mitchell, C.P., 1987. Terminology relating to woody plant biomass and its production. Biomass 12, 79-82.

[10] McAlpine, R., Brown, C., Herrick, A., Ruark, H., 1966. "Silage"sycamore. Forestry Farmer 26 7-16.

[11] Herrick, A.M., Brown, C.L., 1967. A new concept in cellulose production: silage sycamore. Agricultural Science Review 5, 8-13.

[12] Auclair, D., Bouvarel, L., 1992. Intensive or extensive cultivation of short rotation hybrid poplar coppice on forest land. Bioresource Technology 42, 53-59.

[13] Kenney, W., Sennerby-Forsse, L., Layton, P., 1990a. A review of biomass quality research relevant to the use of poplar and willow for energy conversion. Biomass 21, 163-188.

[14] Sims, R.E.H., Senelwa, K., Malava, T., Bullock, T., 1999. Eucalyptus for energy in New Zealand-Part II: coppice performance. Biomass and Bioenergy 17, 333-343.

[15] Grunewald, H., Bohm, C., Quinkenstein, A., Grundmann, P., Jörg Eberts, J., von Wühlisch, G., 2009. Robinia pseudoacacia L.: A Lesser Known Tree Species for Biomass Production. Bioenergy Research 2, 123-133.

[16] Rytter, L., Šlapokas, T., Granhall, U., 1989. Woody biomass and litter production of fertilized grey alder plantations on a low-humified peat bog. Forest Ecology and Management 28, 161-176.

[17] Keoleian, G.A., Volk, T.A., 2005. Renewable energy from willow biomass crops: Life cycle energy, environmental and economic performance. Critical Reviews in Plant Sciences 24, 385-406.

[18] Grislis K, Labrecque L., 2009. Proliferating Willow for Biomass. Silviculture Magazine summer 2009, 12-15.

[19] Mosseler, A., 1990. Hybrid performance and species crossability relationships in willows (Salix). Canadian Journal of Botany 68, 2329-2338.

[20] Kenney, W.A., Sennerby-Forsse, L., Layton, P., 1990b. A review of biomass quality research relevant to the use of poplar and willow for energy conversion. Biomass 21, 163-188.

[21] Labrecque, M., Teodorescu, T., Cogliastro, A., Daigle, S., 1993. Growth patterns and biomass productivity of two Salix species grown under short-rotation intensive culture in southern Quebec. Biomass and Bioenergy 4, 419-425.

[22] Lavoie, F., Savoie, P., D'Amours, L., Joannis, H., 2008. Development and field performance of a willow cuttershredder-baler. Applied Engineering in Agriculture 24, 165-172.

[23] Dickmann, D.I., Kuzovkina, J., 2008. Poplars and willows of the world, with emphasis on silviculturally important species. FAO Rome, Italy

[24] Persson, G., 1995. Willow stand evapotranspiration simulated for Swedish soils. Agricultural Water Management 28, 271-293.

[25] Jackson, M., Attwood, P., 1996. Roots of willow (Salix viminalis L) show marked tolerance to oxygen shortage in flooded soils and in solution culture. Plant and Soil 187, 37-45.

[26] Ledin, S., 1996. Willow wood properties, production and economy. Biomass and Bioenergy 11, 75-83.

[27] Argus, G.W., 1999. Classification of Salix in the New World. Botanical Electronic News 227, http://www.ou.edu/cas/botany-micro/ben227.html.

[28] Labrecque, M., Teodorescu, T., Daigle, S., 1997. Biomass productivity and wood energy of salix species after 2 years growth in SRIC fertilized with wastewater sludge. Biomass and Bioenergy 12, 409-417.

[29] Labrecque, M., Teodorescu, T.I., 2005b. Field performance and biomass production of 12 willow and poplar clones in short-rotation coppice in southern Quebec (Canada). Biomass and Bioenergy 29, 1-9.

[30] Labrecque, M., Teodorescu, T.I., Babeux, P., Cogliastro, A., Daigle, S., 1994. Impact of herbaceous competition and drainage conditions on the early productivity of willows under short-rotation intensive culture. Canadian Journal of Forest Research 24, 493-501.

[31] Defra, 2004. Growing Short Rotation Coppice - Best Practice Guidelines For Applicants to Defra's Energy Crops Scheme. London (UK).

[32] Bergkvist, P., Ledin, S., 1998. Stem biomass yields at different planting designs and spacings in willow coppice systems Biomass and Bioenergy 14, 149-156

[33] Mitchell, C., 1995. New cultural treatments and yield optimization. Biomass and Bioenergy 9, 11-33.

[34] Proe, M., Craig, J., Griffiths, J., 2002. Effects of spacing, species and coppicing on leaf area, light interception and photosynthesis in short rotation forestry Biomass and Bioenergy 23, 315-326

[35] Sennerby-Forsse, L., Ferm, A., Kauppi, A., 1992. Coppicing ability and sustainability. In: Mitchell, C., Ford-Robertson, J., Hinckley, T., Sennerby-Forsse, L. (Eds.), Ecophysiology of short rotation forest crops. Elsevier, London, pp. 146–184.

[36] Willoughby, I., Clay, D.V., 1996. Herbicides for Farm Woodlands and Short Rotation Coppice. London.

[37] Sage, 1999. Weed competition in willow coppice crops: the cause and extent of yield losses. Weed Research 39, 399-411.

[38] Mitchell, C.P., 1992. Ecophysiology of short rotation forest crops. Biomass and Bioenergy 2, 25-37.

[39] Kopp, R.F., White, E.H., Abrahamson, L.P., Nowak, C.A., Zsuffa, L., Burns, K.F., 1993. Willow biomass trials in Central New York State. Biomass and Bioenergy 5, 179-187.

[40] Potter, C.J., 1990. Coppiced trees as energy crops. Final report to ETSU for the DTI on contract ETSU B 1078.

[41] Caslin, B., Finnan, J., McCracken, A.R., 2010. Short Rotation Coppice Willow Best Practice Guidelines.

[42] Hytönen, J., 1995. Ten-year biomass production and stand structure of Salix ['Jaquatica' energy forest plantation in Southern Finland. Biomass and Bioenergy 8, 63-71.

[43] Labrecque, M., Teodorescu, T., 2003. High biomass yields achieved by Salix clones in SRIC following two 3-years coppice rotations on abandoned farmland in southern Quebec, Canada. Biomass and Bioenergy 25, 135-146.

[44] Adegbidi, H., Volk, T., White, E., Abrahamson, L., Briggs, R., Bickelaupt, D., 2001. Biomass and nutrient removal by willow clones in experimental bioenergy plantation in New York State. Biomass and Bioenergy 20, 399-411.

[45] Ericsson, T., 1994. Nutrient cycling in energy forest plantations. Biomass and Bioenergy 6, 115-121.

[46] Alriksson, B., Ledin, S., Seeger, P., 1997. Effect of nitrogen fertilization on growth in a Salix viminalis stand using a response surface experimental design. Scandinavian Journal of Forest Research 12, 321-327.

[47] Ledin, S., Willebrand, E., 1996. Handbook on How to Grow Short Rotation Forests. Swedish University of Agricultural Sciences, Department of Short Rotation Forestry, Uppsala.

[48] Buchholz, T., Volk, T., 2011. Improving the Profitability of Willow Crops—Identifying Opportunities with a Crop Budget Model. BioEnergy Research 4, 85-95.

[49] Perttu, K., 1999. Environmental and hygienic aspects of willow coppice in Sweden. Biomass and Bioenergy 16, 291-297.

[50] Martin, P., Stephens, W., 2006. Willow growth in response to nutrients and moisture on a clay landfill cap soil. I. Growth and biomass production. Bioresource Technology 97, 437-448.

[51] Larsson, S., 2003. Short rotation Willow biomass plantation irrigated and fertilized with wastewaters- Results form a 4-year multidisciplinary field project in Sweden, France, Northern Ireland and Greece supported by the EU-FAIR Programme (FAIR5-CT97-3947) Final Report. Svalov, Sweden.

[52] Jonsson, M., Dimitriou, I., Aronsson, P., Elowson, T., 2006. Treatment of log yard run-off by irrigation of grass and willows. Environmental Pollution 139, 157-166.

[53] Labrecque, M., Teodorescu, T., 2001. Influence of plantation site and wastewater sludge fertilization on the performance and foliar nutrient status of two willow species grown under SRIC in southern Quebec (Canada). Forest Ecology and Management 150, 223-239.

[54] Cavanagh, A., Gasser, M.O., Labrecque, M., 2011. Pig slurry as fertilizer on willow plantation. Biomass and Bioenergy 35, 4165-4173.

[55] Hall, R., Allen, S., Rosier, P., Hopkins, R., 1998. Transpiration from coppiced poplar and willow measured using sap-flow methods. Agricultural and Forest Meteorology 90, 275-290.

[56] Guidi, W., Piccioni, E., Bonari, E., 2008. Evapotranspiration and crop coefficient of poplar and willow short-rotation coppice used as vegetation filter Bioresource Technology 99, 4832-4840.

[57] Aronsson, P., Bergstrom, L., 2001. Nitrate leaching from lysimeter- grown short rotation willow coppice in relation to N- application, irrigation and soil type. Biomass and Bioenergy 21, 155-164.

[58] Carlander, A., Schönning, C., Stenström, T.A., 2009. Energy forest irrigated with wastewater: a comparative microbial risk assessment. Journal of Water and Health 7 413-433

[59] Labrecque, M., Teodorescu, T., Daigle, S., 1995. Effect of wastewater sludge on growth and heavy metal bioaccumulation of two *Salix* species. Plant and Soil 171, 303-316.

[60] Bauer, L.S., 1992. Response of the Imported Willow Leaf Beetle to Bacillus thuringiensis var. san diego on Poplar Willow. Journal of Ivertebrate Pathology 59, 330-331.

[61] Sage, R.B., Tucker, K., 1997. Invertebrates in the canopy of willow and poplar short rotation coppices. Aspects of Applied Biology 49, 105-112.

[62] Blackman, R.L., Eastop, V.F., 1994. Aphids on the world's trees:an identification and information guide.

[63] Collins, C.M., Rosado, R.G., Leather, S.R., 2001. The impact of the aphids Tuberolachnus salignus and Pterocomma salicis on willow trees. Ann. Appl. Biol. 138, 133-140.

[64] Aradottir, G.I., Karp, A., Hanley, S., Shield, I., Woodcock, C.M., Dewhirst, S., Collins, C.M., Leather, S., Harrington, R., 2009. Host selection of the giant willow aphid (Tuberolachnus salignus). Proceedings of the 8th International Symposium on Aphids. REDIA, XCII, , 223-225.

[65] Seago, A., Lingafelter, S.W., 2003. Discovery of Crepidodera Bella Parra (Coleoptera: Chrysomelidae: Alticini) in Maryland and redescription. Journal of the New York Entomological Society 111, 51-56.

[66] Pei, M.H., McCracken, A.R. (Eds.), 2005. Rust Diseases of Willow and Poplar CABI Publishing, CAB International Wallingford, Oxfordshire OX10 8DE UK.

[67] Vujanovic, V., Labrecque, M., 2002. Biodiversity of pathogenic mycobiota in Salix bioenergy plantations, Québec. Canadian Plant Disease Survey 82 138 -139.

[68] Pei, M., Lindegaard, K., Ruiz, C., Bayon, C., 2008. Rust resistance of some varieties and recently bred genotypes of biomass willows. Biomass and Bioenergy 32, 453-459

[69] Parker, S.R., Pei, M.H., Royle, D.J., Hunter, T., Whelan, M.J., 1995. Epidemiology, population dynamics and management of rust diseases in willow energy plantations. Final Report of Project ETSU B/W6/00214/REP. Energy Technology Support Group, Department of Trade and Industry, UK.

[70] McCracken, A.R., Dawson, W.M., 1997. Growing clonal mixtures of willow to reduce effect of Melampsora epitea var. epitea. European Journal of Forest Pathology 27, 319-329.

[71] McCracken, A.R., Dawson, W.M., 2003. Rust disease (Melampsora epited) of willow (Salix spp.) grown as short rotation coppice (SRC) in inter- and intra-species mixtures. Ann. Appl. Biol. 143, 381-393.

[72] Labrecque, M., Teodorescu, T., 2005a. Preliminary evaluation of a living willow (Salix spp.) sound barrier along a highway in Québec, Canada. Journal of Arboriculture 31, 95-98.

[73] Teodorescu, T.I., Guidi, W., Labrecque, M., 2011. The use of non-dormant rods as planting material: A new approach to establishing willow for environmental applications. Ecological Engineering 37, 1430-1433.

[74] Jirjis, R., 1995. Storage and drying of wood fuel. Biomass and Bioenergy 9, 181-190.

[75] Gigler, J.K., van Loon, W.K.P., van den Berg, J.V., Sonneveld, C., Meerdink, G., 2000. Natural wind drying of willow stems. Biomass and Bioenergy 19, 153-163.

[76] Ceulemans, R., McDonald, A., Pereira, J., 1996. A comparison among eucalyptus, poplar and willow characteristics with particular reference to a coppice, growth-modelling approach. Biomass and Bioenergy 11, 215-231.

[77] Mola-Yudego, D., Aronsson, P., 2008. Yield models for commercial willow biomass plantations in Sweden. Biomass and Bioenergy 32, 829-837.

[78] Volk, T., Kiernan, B., Kopp, R., Abrahamson, L., 2001. First and second-rotation yield of willow clones at two sites in New York State. Proceeding of the 5th Biomass Conference of the Americas., Orlando, FL.

[79] Liu, T.T., McConkey, B.G., Ma, Z.Y., Liu, Z.G., Li, X., Cheng, L.L. 2011. Strengths, Weaknessness, Opportunities and Threats Analysis of Bioenergy Production on Marginal Land. Energy Procedia 5: 2378-2386

[80] Licht, L.A., and Isebrands, J.G. 2005. Linking phytoremediated pollutant removal to biomass economic opportunities. Biomass Bioenergy 28: 203-218.

[81] O'Neill, G.J., and Gordon, A.M. 1994. The nitrogen filtering capability of Carolina poplar in artificial riparian zone. J. Env. Quality 23: 1218-1223.

[82] P. Giasson, A. Jaouich, Giasson P. et A. Jaouich. 1998. La phytorestauration des sols contaminés au Québec. Vecteur environnement 31(4):40-53.

[83] Ministère de l'environnement du Québec. 1994. Dix ans de restauration des terrains contaminés (1983-1993). Gouvernement du Québec, Québec, 30 pp.

[84] John Philipsmay, T., Pi J et autline m, and and militar time minimum biology and biotreatment. Longman Scientific & Technical, Singapore Publishers, Singapore.

[85] Vandecasteele, B., E. Meers, P. Vervaeke, B. D. Vos, P. Quataert, and F. M. G. Tack. 2005. Growth and trace metal accumulation of two Salix clones on sediment-derived soils with increasing contamination levels. Chemosphere 58: 995-1002.

[86] Kuzovkina, Y.A., Volk, T.A. 2009. The characterization of willow (*Salix* L.) varieties for use in ecological engineering applications: coordination of structure, function and autecology. Ecological Engineering 35, 1178-1189.

[87] Dowling, D. N., and S. L. Doty. 2009. Improving phytoremediation through biotechnology. Current Opinion in Biotechnology 20: 204-206.

[88] Jiménez, E. M., J. M. Peñalosa, R. Manzano, R. O. C. Ruiz, R. Gamarra, and E. Esteban. 2009. Heavy metals distribution in soils surrounding an abandoned mine in NW Madrid (Spain) and their transference to wild flora. Journal of Hazardous Materials 162: 854-859

[89] Harada, E., A. Hokura, S. Takada, K. Baba, Y. Terada, I. Nakai, and K. Yazaki. 2010. Characterization of Cadmium Accumulation in Willow as a Woody Metal Accumulator

Using Synchrotron Radiation-Based X-Ray Microanalyses. Plant and Cell Physiology 51: 848-853.

[90] Pitre, F. E., Teodorescu T. I., and Labrecque M. 2010. Brownfield Phytoremediation of Heavy Metals using Brassica and Salix supplemented with EDTA: Results of the First Growing Season. Journal of Environmental Science and Engineering 4: 51-59.

[92] Karp, A., Hanley, S. J., Trybush, S. O., Macalpine, W., Pei, M. and Shield, I. (2011), Genetic Improvement of Willow for Bioenergy and Biofuels. Journal of Integrative Plant Biology, 53: 151–165.

[91] Guidi W., Kadri H., Labrecque M. (2012) Establishment techniques to using willow for phytoremediation on a former oil refinery in southern Quebec: achievements and constraints. Chemistry and Ecology 28 (1): 49-64

Chapter 8

Development of Sustainable Willow Short Rotation Forestry in Northern Europe

Theo Verwijst, Anneli Lundkvist, Stina Edelfeldt and Johannes Albertsson

Additional information is available at the end of the chapter

1. Introduction

Modern willow short rotation forestry is based on traditional woodland management which uses the ability of certain tree species to grow new shoots from the stump after being cut down. Depending on site fertility, growing season length, initial planting density and species, willows may be coppiced from once a year to every fifth year, and the stands may remain productive over several decades. Traditionally, small-scale willow plantations have been used for fuel, fodder, convenience wood, basked making, bee keeping, and for horticultural purposes. Willows also may be used for erosion control, including wind and water erosion, and to avoid snow drift along roads. While the traditional use of willow is declining rapidly in Europe, the use of willow as an alternative crop for farmers has led to an increasing interest in willow breeding and cultivation [1]. A renewed research effort on short rotation willow coppice plantations in Sweden commenced in the late 1960's due to a predicted shortage of raw materials for the pulp and paper industries, which turned out to be a false alarm. However, the 1970's energy crisis constituted a new driver to continue research on willows as a source of biomass for energy purposes. Additional drivers, such as employment issues in the Swedish country side, and environmental concerns also influenced research funding rates and directions towards willow short rotation coppice. In the late 1980's willow growing for energy was implemented at a larger scale and commercialized in Sweden. A tax on carbon dioxide emissions for the combustion of fossil fuel in heat production was introduced by the Swedish government during the 1990's and created more favorable market conditions for investment in and implementation of biofuel systems [2]. In 1996, Sweden joined the European Union, which employed an agricultural policy in which subsidy levels to farmers constantly were altered and adapted to short term market situations. As willow growing is a long term commitment which requires longer term investments, this EU-policy promoted the use of annual crops, and the exponential increase of areas under willow cultivation leveled out after 1996 and even started to decline.

In the meantime, the Swedish concept of large-scale willow cultivation for bioenergy purposes was exported to several EU-countries, notably to the UK and Poland, and a development of similar growing systems also was pursued overseas, in New Zealand and in the USA [3].

It was recognized early that willow growth concurs with potentially high evapotranspiration rates [4] and high nitrogen retention rates [5]. Willow species also may exhibit selective uptake of heavy metals [6], which underlies the potential to use willow as a phytoextractor for e.g. Cd from polluted soils [7]. These special traits of willow have allowed a further development of short-rotation willow coppice systems for environmental purposes [8]. Willow growing systems may be used as vegetation filters for purification of waste water [9], for cleaning of polluted drainage water from agricultural land [10] and as a recipient of nutrients from municipal sludge [11]. As willow stands are harvested at regular intervals, the pollutants are removed from the soil-plant system, while added nutrients and water enhance the systems' biomass production. These systems then function as multi-purpose systems, simultaneously aiming at biomass production for energy purposes and provision of environmental services, while producing clean water and neutralizing potentially hazardous compounds. Several efforts have been made to assess the economic gains of such multi-purpose systems [e.g. 12, 13], and Volk et al. [3] concluded that the economic valuation of the environmental benefits is necessary for a further deployment of woody crops.

In the following sections, a brief overview will be given of the plant material and growing system used in willow short- rotation forestry (SRF) and of the history of willow research, with a focus on the developments in Sweden. We then continue with a description of the development and implementation of willow SRF in commercial practice, and with the current guidelines for commercial willow growing. We also present an update of recent research, performed to improve the productivity and sustainability of willow short rotation forestry as an agricultural crop for bioenergy purposes, and include some results of ongoing research projects.

2. Species characteristics and natural distribution of willows

The genus *Salix* comprises about 350 to 500 different species worldwide [14] and is taxonomically complex and difficult to arrange in distinct sub-groups, probably due to intersectional and intersubgeneric polyploidy [15]. About 10% of the willow species consist of deciduous tree species, some of which may attain a height of > 20 meter. However, the vast majority consists of multiple stemmed trees and shrubs, and also a number of very short procumbent species can be found, not exceeding the height of the herb-layer in which they reside. Willow mainly is a boreal-arctic genus, with its natural distribution primarily in the northern hemisphere. Most willow species are found in China and in the former Soviet Union, and some indigenous species are present in India and Japan. The genus also occurs naturally in the southern hemisphere in Africa and in Central- and South America [14], and has been introduced in Australasia and New Zealand. Many species have been transferred beyond their natural range. The short rotation coppice systems currently in use in Sweden are mainly based on *Salix viminalis*, which was introduced in the 1700's from continental Europe for the purpose of basket making, and on their hybrids with *S. burjatica* and *S. schwerinii*, recently introduced from Siberia.

Early records of willow cultivation date from 2000 years ago in the Roman Empire and in modern times willow breeding and selection programs have been recorded from Sweden, the UK, Belgium, France, Croatia, Poland, Hungary, former Yugoslavia, Romania, Bulgaria and China, but also outside Eurasia in New Zealand, Argentina, Chile, Canada and in the USA. The development of molecular methods in plant breeding is likely to speed up the selection of new and viable material [16] and is envisaged to lead to a willow crop which is less prone to pests and diseases and which can be managed with lower inputs than the current systems [17].

The widespread interest in the willow genus is due to the fact that many of its species, which are light demanding pioneer trees, exhibit a very high growth rate in their juvenile stage. Many willow species can easily be propagated by means of cuttings, and most species and their hybrids will generate new shoots abundantly after cutting down older shoots and stems [18]. Under Swedish conditions, willow has a very high and well documented growth potential [19] which, though, is not completely realized in commercial short rotation forestry [20]. To fully exploit the growth potential of willows, a soil fertility level is required which is comparable with those found on conventional agricultural soils in Sweden. To maintain growth in the long term, dry sites have to be avoided and nutrients have to be added at a rate which balances nutrient removal by harvest. Compared to conventional forestry, willows require a relative intensive management, but compared to conventional agricultural practice, management input is much lower.

3. Growing systems & population dynamics

Given the huge range in size, growth form and coppice ability in the willow genus, production systems for willow may vary from single-stemmed systems with less than 500 trees ha^{-1} and a rotation period of over 20 years, to systems which contain over 4×10^4 plants plant which generate over half a million shoots ha^{-1} in a one-year coppice cycle. In the remainder of this chapter, we focus on growing systems which are generated from cuttings, at a planting density of 1×10^4 to 1.5×10^4 cuttings ha^{-1}, and treated as a coppice system, undergoing multiple cutting cycles. In Scandinavian conditions, one season may be too short to replenish carbohydrate reserves in willow stubs after harvest, and a one-year harvest cycle may deplete a plantation and compromise its viability [21]. Cutting cycle lengths in Swedish practice have been 3 to 5 years, and with the introduction of faster growing clones, cutting cycle lengths now are being decreased to 2 to 4 years. In commercial practice, a double row system is employed (Figure 1). However, Bergkvist and Ledin [22] showed that planting design could be adjusted, within certain limits without losing yield potential, to the requirements of tractors and machines used in managing the *Salix* stand.

Figure 1. Machine planting of willow by means of a Woodpecker 601, using long rods and planting three double rows at a time (Photo: Nils-Erik Nordh).

The development of a population of willow stems is constrained by competitive interactions which lead to self-thinning, yield-density effects and to skewed size-frequency distributions of stems [23, 24]. Those effects of competitive interactions need to be accounted for when determining optimal plant spacing and harvest frequency. Especially in dense willow coppice, not only shoot mortality but also extensive stool mortality may occur [25], thereby leading to lasting gaps and production losses [26]. Studies on the long term dynamics of willow coppice have shown that an initial variability in plant size becomes enlarged over time, that self-thinning leads to mortality of the initially smallest stools [27], and that the competitive hierarchy between stools is preserved over harvest [26]. As soil factors are known to be important determinants of willow growth [28, 29, 30, 31], differences in soil at field scale likely underlie the initial size variability between plants. Differences in cutting quality also may cause an initial variability in growth performance between plants (see section 4.2). To be able to detect possible effects of cutting quality and to separate those from soil factors, it is advantageous to perform controlled experiments which allow the relative variation to be attributed to only a few factors. Verwijst et al. [32] compared the relative variation in shoot height of willow populations grown in the field with the relative variation of populations grown in boxes which had a standard soil and were treated as similar as possible with regard to fertilization and irrigation (Figure 2). The controlled experiments showed a decreased relative variation and enhanced the detection of cutting quality traits with relevance for early establishment success.

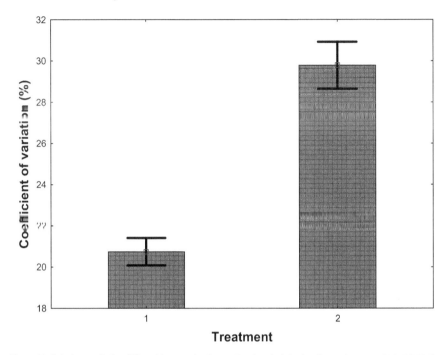

Treatment

Figure 2. Relative variation (%) and its standard error in plant height for shoots from cuttings planted in a controlled environment (Treatment 1) versus shoots planted in the field (Treatment 2).

As willow is a relatively new crop, advances in willow breeding generate a steady increase in potential and attainable yield [33, 34]. This increase in biomass yield is estimated to be 50 to 100% since the 1970s'. This means that spacing, harvest frequency and fertilization have to be adapted to the rapidly evolving new plant material, in order to avoid mortality and ensure a high productivity also during the later cutting cycles. Most of the planted willow stands in Sweden consist of monoclonal stands or blocks of monoclonal units. However, such monoclonal stands are vulnerable to pathogen adaptations [35] and it has been shown that clone mixtures may be effective against the spread of diseases [36]. However, the relative competitive power of willow clones does differ, which means that certain clones may be outcompeted by other ones in mixtures of clones. If a mixture consists of only a few clones and one of the components is attacked by a pathogen, the susceptible clone is likely to be outcompeted by the others, thereby causing gaps, a delayed stand closure and lower productivity in later cutting cycles [37]. Furthermore, as clone-site interactions have been reported for willow, and the performance of clones in mixtures can not be predicted from their performance in pure stands [24], successful clone combinations are expected to be highly site specific.

3.1. Site choice and preparation

Many willow species do have abroad ecological amplitude. However, to obtain a high productivity, willow has specific site requirements. Being a pioneer species, willow is light demanding, and a rapid establishment can only be achieved without competition by weeds for light. Once established, the leaf area index of a willow canopy will exceed 6 $m^2 \times m^{-2}$ [4, 38] and will suppress weed growth. Willow thrives on most agricultural soils, as long as the pH is in the range of 5 to 7 [39]. Water use efficiency of a willow crop is about 4 to 6 $g \times kg^{-1}$ [4]. This is a high value compared to values of other tree species, but given the potentially high biomass production of willow, water availability is conceived as a critical factor in willow SRF [40]. Consequently, lighter soils, especially in drier areas, should be avoided for willow growing. A low precipitation during the growing season can be compensated for if winter precipitation is abundant and soils have a good water holding capacity or do have access to groundwater. While many willow species have a boreal-arctic origin and are native to northern temperate regions, fast-growing hybrids may be susceptible to frost damage from bud-burst and onwards. If planted at frost-exposed sites, a single night frost may decrease a single year's productivity by 50% [41] and will also impact negatively on the biomass production in the following years. Therefore, sites prone to late spring frost should be avoided and it is important to choose clones which have a site-adapted phenology with regard to timing of bud burst. Willow can be harvested with a reasonable cost-efficiency on sites which are 5 ha or larger, and even on slightly smaller sites if willow is harvested on adjacent sites. Planting and harvest equipment for willow requires a relatively widely spaced headland (10 to 12 m in width), which means that single willow fields should not be smaller than 2 ha, and easily could be reached by the harvest machines [42]. Larger stones also should be removed from the soil surface, as they may damage harvest equipment. As planting (see section 4.2) requires a well prepared seed bed, autumn plowing and early spring seedbed preparation are common measures prior to planting. Such preparation has to go along with adequate weed control (see section 4.3). Another selection criterion for willow growing sites is the proximity to a consumer, usually a combined heat and power plant. As moist willow chips do have low energy content per volume, transportation distances by road should be minimized [13]. Finally, willow growing is a form of land-use, and as such, it may interfere with a range of other interests than sheer biomass production. Short rotation forests may affect landscape views, the environment and biodiversity in a positive or negative way, depending on the functions that we require from a semi-natural landscape element, and on how we choose to integrate such functions in a single growing system [1, 44].

3.2. Planting & cutting quality

One of the large advantages of most willows is that they can be propagated vegetatively by means of cuttings. Traditionally, cuttings of about 20 cm in length were produced manually from 1-year old long rods. These cuttings were taken during the winter period, when willow is dormant, and could be stored in a fridge until planting in spring. During commercialization of the growing system in Sweden in the late 1980s, manual planting was replaced by machine planting. Establishment costs for short rotation willow coppice decreased substantially during the initial phase of commercialization in Sweden [45]. This was mainly achieved by mechanisation of planting, employing equipment which, in one process, cuts willow rods (1.8 – 2.4 m. long) into cuttings and then plants them (Figure 1). These cuttings are around 18 to 20 cm long, and the cutting is pressed down into the prepared soil so that only 1-2 cm protrudes above soil surface. This is believed to provide the cutting with good soil contact, thereby minimizing the risk of drying out [46]. Field storage of cuttings can result in water loss and reduce shoot survival and biomass production. This problem has partly been overcome by the use of entire shoots, which are considered to be more resistant to desiccation than cuttings [47]. Volk et al. [18] also pointed to risks of desiccation and showed that a prolonged time of field storage after cold storage may lead to a decrease in survival and growth rate.

Stage	Description
1	No sign of bud swelling, the tip of the bud is tightly pressed to the shoot.
2	The tip of the bud starts to bend from the stem, bud scales are starting to open and the length of the shoot tip is 1–4 mm.
3	The shoot tip is 5 mm or longer, protruding leaves are put together.
4	New leaves start to bend from each other.
5	One or more new leaves are perpendicular to the shoot axis.

Table 1. Assessment criteria for bud burst stages.

Cutting size (length and diameter) has positive effects on subsequent willow growth. The positive effects of cutting size on growth and survival decline with increasing sizes ([49, 50, 51], and Rossi [52] found that the differences in cutting length with relevance for establishment in practice are to be found between lengths of 10 and 20 cm. Positive effects of cutting size generally are attributed to the size of the carbohydrate pool available for allocation to roots and shoots [53]. The effect of cutting length may also be associated to the ability of longer cuttings to withstand soil desiccation [54]. The phenological development of buds and shoots is affected by cutting size and also by the height above ground from where the cuttings were taken [51]. Using the simple assessment criteria for bud development as described in Table 1, bud development, a few weeks after planting, is a function of the diameter size of the planted cutting (Figure 3). However, cuttings derived from apical positions along shoots display for a given diameter a higher shoot biomass production than cuttings derived from the more basal parts (Figure 4). As willow rods display a taper, the question arises which of the two factors (cutting size or position) is the strongest determinant of shoot biomass production during early establishment.

A further evaluation of produced shoot biomass on the cuttings showed that cutting size by far is the single most important determinant of early biomass production, which led to the recommendation to employ thicker cuttings and to discard the thinner apical parts from long rods. While the introduction of planting machines has increased the speed of planting and reduced planting costs, ongoing research indicates that planting machines may cause damage to cuttings, especially when planted in compacted soils. Preliminary results by Verwijst et al. [32] and by Edelfeldt et al. [55], suggest that that undamaged cuttings had a better growth performance than visibly damaged cuttings. Planting by machine on hard soil resulted in a relatively large number of cuttings landing on the soils surface. Soil compaction and machine planting interacted with cutting dimensions, the poorer performance of thinner cuttings being more pronounced in compacted soil (Figure 5).

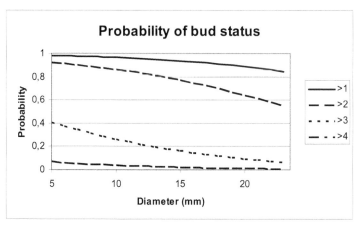

Figure 3. Probability of bud status (see Table 1) at average values for five clones and cuttings derived from a position of 95 cm above soil surface, a few weeks after planting. Probability of high bud status decreases with diameter.

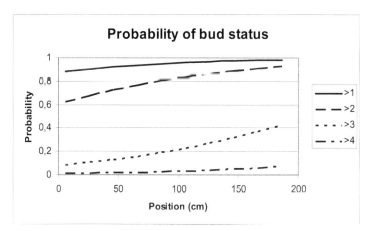

Figure 4. Probability of bud status (see Table 1) at average values for five clones and diameter 12.5 mm, a few weeks after planting. Probability of high bud status increases with the original height position of the cutting along the rods from which is was derived.

Furthermore, machine planting also increased the relative variation of shoot height (Figure 6) compared to hand prepared and planted cuttings. Consequently, to obtain a faster and more even establishment of willows, Edelfeldt et al. [55] recommend thorough soil cultivation prior to planting, further development of planting machines to minimise damage to cuttings at planting, and the use of cuttings with a diameter of at least 10-11 mm.

Figure 5. Cuttings planted by machine in a hard soil were transformed to a soft soil to isolate the effect of machine planting from other factors. The thinner cuttings were visually damaged and displayed a lower sprouting performance than the thicker ones (Photo: Nils-Erik Nordh).

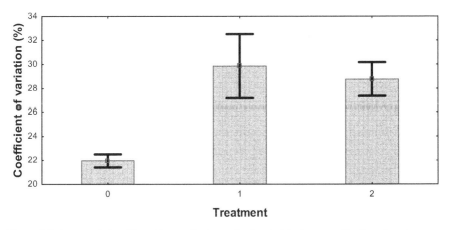

Figure 6. Relative variation (%) and its standard error in plant height for manually planted cuttings (Treatment 0) versus machine planted cuttings in Soft and Hard soil (Treatments 1 and 2, respectively).

3.3. Weed control

Weed control is necessary when establishing willows from cuttings, because its takes a relatively long time for willow cuttings to develop attributes which make them competitive against weeds. Competition is an interaction between plants which require the same limited resources like nutrients, water and light. Harper [56] defines competition as 'An interaction between individuals brought about by a shared requirement for a resource in limited supply and leading to a reduction in the survivorship, growth and/or reproduction of the individuals concerned', and thereby points to the effects of competition. The aim of weed control is to ensure that as much resources as possible are accessible for the crop and not for the weeds, and to reduce or delay growth and development of the weed flora [57].

Willows under establishment from cuttings have a relatively low competitive power against weeds because it takes a while for willows to develop roots needed for the uptake of nutrients and water. Consequently, perennial weeds, which have a developed root system prior to the onset of leaves, have to be removed completely before planting willows. This commonly is done by means of one or two applications of a glyphosate-based herbicide, applied at the appropriate rate, during the summer/autumn prior to spring planting. If the area has not been used for agricultural purposes for a number of years before planting, it is recommended to grow cereals there for at least one season to ensure an adequate weed control [42]. The relative competitive ability is also affected by seed rate (plant density), which is low for willow (between 1 and 2 cuttings m^{-2}), in comparison to the amount of germinating annual weeds triggered by seed bed preparation. Such weeds may germinate only a few days after seed bed preparation, while it may take a week or more for willow cuttings to exhibit a first bud burst after planting. This implies that the time between the last seed bed preparation and willow planting should be minimized. To counteract the effects of the inherent differences in relative emergence time between willow and weeds, soil cultivation by different types of cultivators, rototillers or harrows are recommended as a weed control measure during willow establishment [46, 58]. There are also different soil-applied herbicides that are permitted to be used at planting or shortly thereafter. Given the low planting density of willow cuttings, a full canopy closure, which for willow implies a leaf area index > 6 m^2×m^{-2} [4, 38] is hardly ever reached during the establishment year, which means that if weeds are not kept back during the establishment year, they may establish and compete with willows for light. The use of mechanical weeding may therefore proceed even after bud burst and early shoot formation in willows. As the cuttings are well fixed in the soil and young willow shoots are flexible, they will not become damaged by this treatment. The current recommendation is to perform these control measures at least three times during the first year [46].

Weed control might also be necessary to perform the year after planting depending on weed management success the first year, clones and site conditions. As the willow plants will be better established by then, it is usually enough to perform mechanical weeding two times early in the season [46]. Another possibility is to spray a soil-applied herbicide well before bud burst [42] or to use a selective herbicide during spring or early summer [46]. Weed control the second year usually requires that the first year shoots are cut back. This practice has been questioned [59] and is no longer recommended in Sweden [42]. If weed control has been efficient during the establishment phase, no additional measures are required to control the weeds the following years. If early plant mortality has led to gaps in the stand, weeds may establish and maintain themselves below canopy gaps (Figure 7). In case weeds survived below such gaps, weeds may be controlled directly after each harvest.

If the weeds are not controlled during the establishment phase, willow growth might be dramatically reduced. Field experiments conducted in Southern Sweden by Albertsson in 2010-2012, with 10 modern willow varieties, grown both with- and without weeds, have shown that weeds can increase plant mortality, and reduce growth the first year by more than 95% [42], see Figure 8. Several other studies have also shown that willow, in the establishment phase, is very sensitive to competition from other plants [60, 61, 62]. Preliminary data from the Swedish study also suggest that there is an interaction between voles and weediness, since plots with weeds show more damage by voles than plots without weeds, thereby making weed control even more important.

Figure 7. Poor establishment of willow leads to gaps in which weeds may establish, thereby causing the need for prolonged weed control after a first harvest (Photo: Nils-Erik Nordh).

Weeds in willow short rotation coppice might, in the future, be controlled with other measures than the above mentioned. Studies are ongoing to investigate if willow clones differ in their ability to compete with weeds. Fast initial growth, early bud burst, fast canopy closure and the ability to tolerate or release allelophatic substances might be favorable weed competing traits. If differences exist, it might be possible to breed for these traits or to use competitive willow varieties that combine well with a specific weed control measure.

Different cover crops such as rye (*Secale cereale* L.), dutch white clover (*Trifolium repens* L.), buckwheat (*Fagopyrum esculentum* Moench) and caragana (*Caragana arborescens* Lam.) have been studied as a way of controlling weeds and improve nutritional management in willow [63, 64, 65]. However, there is still more research to be conducted in this area before a suitable willow cover crop system is ready for commercial use. Mechanical weeding techniques are under constant development and recent results indicate that automatic intra-row weeding is possible [66]. Hence, these techniques may be further developed to be used in willow since weeds within the rows are hard to control mechanically with conventional equipment.

Figure 8. Weeds were removed mechanically and by hand in the willow stand to the left while no weed control measures were performed in the willow stand to the right. The photo was taken five months after planting (Photo: Johannes Albertsson).

3.4. Fertilization

Most field-based cropping systems do have an actual production which is well below their potential production. The potential production of a crop is determined genetically by its nutrient-, water- and light use efficiency. But given those efficiencies, a field environment hardly ever constantly provides optimal supply of water, nutrients and light to the crops. The production which is attained after restriction by abiotic factors such as light, water and nutrients is called attainable production, and can be regulated by site choice and fertilization. Actual production is usually lower than the level of attainable production, being utterly restricted by the effects of biotic agents, such as herbivores and pathogens.

Consequently, plant breeding and selection partly strive to generate plant material with a high resistance against pests and diseases, but also to generate material with a positive response for treatments such as fertilization. From a farmers' perspective, fertilization may be applied if it enhances profitability of the cropping system. Profitability then is a function of costs for fertilizers, the net value of the crop, and of the fertilization effect, i.e. the additional biomass increment per unit added fertilizer. The willow clones that have been released during the last decades in Sweden display a higher actual productivity that the earlier ones [45, 67], and this seems amongst others to be the result of a clonal selection towards a higher shoot/root allocation patterns, resulting in a higher harvestable biomass increment per unit fertilizer. While selection thus promotes a positive response to fertilization and irrigation, it may also increase the susceptibility of clones for incidental drought periods [68].

Recommendations for farmers with regard to fertilization of willow coppice on agricultural land during the last decades in Sweden have been subject to a great deal of confusion, due to the fact that fertilizer costs, net crop revenue and fertilization response of the crop all rapidly have been changing through time. Early recommendations by Ledin [69] were based on fertilization trials with older willow clones and on economic calculations which accounted for projected crop values which were not met by the market. Net values for different fertilization strategies under different scenarios with regard to fertilization costs and actual net crop values recently have been calculated [70] after field based parameterizations of the fertilization response of more recent willow clones. It was found that fertilization responses differed widely between clones and sites and that fertilization should be adapted to the local conditions. Under current market conditions and using recently released willow clones, fertilization can greatly enhance profitability. The need for fertilization of modern clones in a first cutting cycle could not be assessed due to lack of data. However, fertilization during the first year may positively affect weed growth, and is therefore not recommended. Plantations with modern willow clones should be fertilized with at least 220 kg N ha^{-1} during the second and consecutive cutting cycles. Annual fertilization in willow stand would require a further machine development, as conventional machinery cannot enter tall willow stands.

Fertilization may also be performed with nutrient-rich residues such as municipal wastewater and sludge to willow short rotation coppice [71] and may render a more cost-effective and sustainable cultivation. Rosenqvist and Ness [72] provide an economic analysis of leached purification through willow coppice vegetation filters and showed that economic gains were made compared to conventional purification, while an increased biomass production led to additional economic gains. It also is concluded that willow vegetation filters are more cost-effective than conventional treatment methods and may facilitate recycling of valuable products in society [5]. This conclusion is sustained by other assessments of the economic gains of such multi-purpose systems [e.g. 12, 13]. Volk et al [3] even concluded that the economic valuation of the environmental benefits is necessary for a further deployment of woody crops.

3.5. Control of pests and diseases

Attainable biomass production of a crop, as determined by its genetics and actual resource levels provided in a particular field situation, is usually reduced by the action of pathogens and herbivores. Especially in genera with species that hybridize easily, such as willow, the relationship between plant breeding and pest and disease control is strong, because such genera in general attract many kinds of insects and pathogens. Plants may be well adapted to a specified range of abiotic conditions, which display a site specific variation. However, pests and diseases are biotic factors which not only vary in space and time, but may also co-evolve with plants. Consequently, potentially pathogenic organisms may be present and may do little harm for longer periods in a willow stand, until virulent strains develop which may be very clone specific. For instance, susceptibility to defined pathotypes of leaf rust (*Melampsora epitea*) is rather clone specific [73]. Consequently, it is important that new clones are released constantly by breeding programs and that a broad genetic base is used, targeting a broad tolerance to a range of pathogens. Poplar breeding programs in Western Europe previously have underrated this issue, resulting in the destruction of many poplar stands by leaf rust varieties that managed to adapt to the poplar clones [74]. In willow breeding, this issue was acknowledged early. Development of new high producing willow clones was initiated in Sweden in 1987 by Svalöf-Weibull AB [33]. The main purpose of the breeding program was to develop high yielding clones resistant towards pests, frost, and diseases, and with morphology suitable for mechanical harvesting. From 1996 to 2002 several new clones were developed in cooperation between Svalöf-Weibull and Long-Ashton research in UK, also with a strong focus on pest and disease resistance [34]. Strong advances were made early with regard to leaf rust in willow [75] and resistance of willow to several insect species has also been exploited [76, 77]. Production losses between 20 and 40% have been recorded in willow after defoliation by insects [78]. Willow, however, usually recovers well after defoliation, and as the population dynamics of many insects is erratic, and under control of very many factors, damage prevention by means of breeding towards resistance has been chosen, instead of the use of pesticides. Salix has probably the best environmental profile among the arable bioenergy crops available today, partly because neither fungicides nor insecticides are used in the production. This environmental profile is largely an outcome of plant breeding because resistance to pests and diseases, such as leaf rust and certain insects, has been highly prioritized since commercial breeding started in Sweden 25 years ago [79, 80].

3.6. Harvest and logistics

During early commercialization of the willow coppice system as an agricultural crop in Sweden, funding agencies made the decision to put the far majority of the development costs for harvest machines on the account of commercial machine developers. This resulted in a situation in the early 1990s where many willow stands needed to be harvested before self-thinning would lead to an irreversible mortality among willow stools and long-term production losses, while harvest machines still had to be developed and assembled. This is one of the reasons for the early commercial yields to be disappointingly low (see section 4.7).

Fortunately, a variety of willow harvest machines are on the market now, and recent technical improvements greatly enhance harvesting speed while lowering the costs for willow harvesting. In Sweden, willow is usually harvested during the winter, when the soil is able to carry heavy machinery and when willow chips can be transported to district heating plants for direct use, without long-term storage (Figure 9).

Figure 9. Willow harvest by means of a self-propelled chipper which blows the willow chips in an adjacent container (Photo: Nils-Erik Nordh).

However, mild and wet winters may prohibit the use of heavy harvesters, which means that either lighter equipment has to be developed or that the harvest season has to be extended. Expanding the harvesting season for willow biomass crops would expand the time period over which it can be a part of the fuel supply and increase the number of acres that a single harvesting machine could cover in a single year. This would likely increase the demand for willow and certainly reduce harvesting costs, because capital expenditures for a harvester would be spread across more tons of biomass. Nordh [81] investigated the possibility to extend the harvest season, focusing on the re-growth capacity of willow coppice after harvesting , and found that willow (clone Tora) could be harvested from autumn, prior to the onset of dormancy, until late spring, when bud burst already had commenced. Early and late harvest did not increase plant mortality, but it could result in a slight production decrease in the consecutive season.

Apart from direct chipping (Figure 9), willow biomass can be baled (Figure 10) and fragmented in a later stage, possibly after storage, which will decrease moisture content of the willow biomass.

Figure 10. Willow harvest may be performed by means of a machine which produces bales that can be transported by conventional machines. Bales may be stored to obtain biomass with lower moisture content (Photo: Nils-Erik Nordh).

To harvest willow rods for conventional planting by means of a machine, equipment has been developed which can harvest entire one-year old shoots. Mature stands can also be harvested by means of a whole-shoot harvester (Figure 11) which may carry its load to the headland for further transportation. Special equipment has been developed to make bundles from a pile of whole shoots, thereby improving further transportation logistics. As willow is a low-density fuel, willow should preferably be cultivated in the proximity of the consumer, to decrease transportation distances and costs.

3.7. Yield levels

Biomass productivity of short rotation coppice has been studied for several fast growing species in many places of the world, showing an average annual production of 10 to 20 oven dry tonnes (odt) ha^{-1} in most places [82]. In intensively irrigated and fertilized willow plots

Figure 11. A tractor-pulled whole shoot harvester, unloading willow shoots at the headland (Photo: Nils-Erik Nordh).

In southern Sweden, growth rates of ≥ 30 odt ha⁻¹ yr⁻¹ have been recorded [83] The potential production of a certain genotype can only be reached if resources (light, water and nutrients) are permanent available and without limitations, and in the absence of pests and diseases. An analysis of short rotation coppice yields in Sweden over the period 1989-2005 showed disappointingly low mean annual production figures of 2.6, 4.2 and 4.5 odt ha⁻¹ during the first, second and third cutting cycles, respectively [20]. These low figures can partly be explained by the use of old clones, which have a much lower potential production than those which were released later [34] and which have a relatively high susceptibility to pathogens. Other reasons for this low productivity are site choice, as farmers have been reluctant to use the better soils for willow plantations, and a very poor management. Many of the early plantations never received fertilizer and suffered from a poor establishment due to inadequate weed control. However, annual average yields over 10 odt ha⁻¹ have been reached in commercial plantations if fertilization was applied and adequate weed control performed [84], and did not require more than an average availability of water. Taking account of the water use efficiency of willow and precipitation during the growing season, Lindroth & Båth [85] calculated the annual maximum yield to be 8–9 odt ha⁻¹ for north-eastern, 9–10 odt ha⁻¹ for eastern and 11-17 odt ha⁻¹ for southern and south-western Sweden. Studies confined to the new willow clones which have been developed in cooperation between Svalöf-Weibull and Long-Ashton research in UK between 1996 and 2002 confirm that willow breeding has been leading to higher yields in commercial practice. For the new clones, reported yields vary between 5 and 12 odt ha[-1], with extremes between 2 and 18 odt ha[-1] yr[-1] [34, 86, 87, 88]. This large variation seems to be related to interactions between clones and sites [33, 89].

4. Conclusion

Willow short rotation coppice systems are relatively new as a farm crop and both farmers and extension workers in Sweden have gone through a learning process which is now leading to higher yields in commercial plantations. Traditional willow breeding and selection are already greatly contributing to increasing yields, and it is expected that future improvements of the willow varieties will result in a significant increase of the yields in the near future. Many of the early field research results are currently extended with more controlled experiments, and help to improve short rotation coppice management. Although the early commercial implementation of willow coppice did not meet the expectations with regard to yield, profitability and areal expansion of willow coppice, analyses of the early commercial fields contribute to the improvement of stand management, and of the planting, harvest and transport logistics. Further developments of willow coppice as multi-purpose systems, including environmental functions, are promising. Current research suggests that there is room for further improvements with regard to cutting quality, planting, weed control and fertilization, all of which will contribute to higher future yields.

Author details

Theo Verwijst*, Anneli Lundkvist and Stina Edelfeldt
Department of Crop Production Ecology, Swedish University of Agricultural Sciences, Uppsala, Sweden

Johannes Albertsson
Department of Plant Breeding and Biotechnology, Swedish University of Agricultural Sciences, Alnarp, Sweden

Acknowledgement

We kindly acknowledge the financial support from The Swedish Research Council for Environment, Agricultural Sciences and Spatial Planning, Stockholm, Sweden; The Swedish University of Agricultural Sciences (SLU), and The Thermal Engineering Research Association (Värmeforsk), Sweden. We thank Nils-Erik Nordh for many of the photographs which illustrate this chapter. Inger Åhman, Nils-Ove Bertholdsson, David Hansson, Sten Segerslätt, Gunnar Henriksson, Stig Larsson, Gabriele Engqvist, Bertil Christensson and Sven Erik Svensson all are acknowledged for their advice and constructive co-operation in

* Corresponding Author

the different phases of our willow work. Erik Rasmusson, Eskil Kemphe, Fatih Mohammad, Vehbo Hot, Ingegerd Nilsson, Nils-Erik Nordh and Richard Childs are kindly acknowledged for practical help with the experiments. Finally we thank all the agriculturally skilled and hard working students that have helped us coping with all the experiments through the years.

5. References

[1] Verwijst T (2001) Willows: An Underestimated Resource for Environment and Society. The forestry chronicle 77: 281-285.

[2] McCormick K, Kåberger T (2005) "Exploring a Pioneering Bioenergy System. The Case of Enköping in Sweden". Journal of cleaner production 13: 1004-1005.

[3] Volk TA, Verwijst T, Tharakan PJ, Abrahamson LP, White EH (2004a) Growing Fuel: A Sustainability Assessment of Willow Biomass Crops. Frontiers in Ecology and the Environment 2: 411–418.

[4] Lindroth A, Verwijst T, Halldin, S (1994) Water-use efficiency of willow: variation with season, humidity and biomass allocation. Journal of hydrology 156: 1-19.

[5] Aronsson P, Perttu K (2001) Willow Vegetation Filters for Wastewater Treatment and Soil Remediation combined with Biomass Production. The forestry chronicle, 77: 293-299.

[6] Landberg T, Greger M (2002) Interclonal Variation of Heavy Metal Interactions in *Salix viminalis*. Environmental toxicology and chemistry 21: 2669-2674.

[7] Klang-Westin E, Eriksson J (2003) Potential of *Salix* as a Phytoextractor for Cd and Moderately Contaminated Soils. Plant and soil 249: 127-137.

[8] Mirck J, Isebrands JG, Verwijst T, Ledin S (2005) Development of Short Rotation Willow Coppice Systems for Environmental Purposes in Sweden. Biomass & bioenergy 28: 219-220.

[9] Perttu KL, Kowalik PJ (1997) '*Salix* Vegetation Filters for Purification of Water and Soils.' Biomass & bioenergy 12: 9-19.

[10] Elowson S (1999) Willow as a Vegetation Filter for Cleaning of Polluted Drainage Water from Agricultural Land. Biomass & bioenergy 16: 281-290.

[11] Perttu KL (1999) Environmental and Hygienic Aspects of Willow Coppice in Sweden. Biomass & bioenergy 16: 291-297.

[12] Rosenqvist H, Aronsson P, Hasselgren K, Perttu K (1997) Economics of Using Municipal Wastewater Irrigation of Willow Coppice Crops. Biomass & bioenergy 12: 1-8.

[13] Lewandowski I, Schmidt U, Londo M, Faaij A (2006) The Economic Value of the Phytoremediation Function – Assessed by the Example of Cadmium Remediation by Willow (*Salix* ssp). Agricultural systems 89: 68-89.

[14] Argus GW (1999) Classification of *Salix* in the New World. Botanical Electronic News 227. 6p.

[15] Argus GW (1997) Infrageneric Classification of *Salix* (*Salicaceae*) in the New World. Systematic Botany Monographs 52. 121 p.

[16] Kopp RF, Smart LB, Maynard CA, Isebrands JG, Tuskan GA, Abrahamson LP (2001) The Development of Improved Willow Clones for Eastern North America. The forestry chronicle 77: 287-292.

[17] Weih M, Bonosi L, Ghelardini L, Rönnberg-Wästljung AC (2011) Optimizing Nitrogen Economy under Drought: Increased Leaf Nitrogen is an Acclimation to Water Stress in Willow (*Salix* spp.). Annals of botany 108: 1347-1353.

[18] Mitchell CP, Ford-Robertson JB, Hinckley T, Sennerby-Forsse L (Editors) (1992) Ecophysiology of Short Rotation Forest Crops. Elsevier, London.

[19] Christersson L (1986) High Technology Biomass Production by *Salix* Clones on a Sandy Soil in Southern Sweden. Tree Physiology 2: 261-277.

[20] Mola Yudego B, Aronsson P (2008) Yield Models for Commercial Willow Biomass Plantations in Sweden. Biomass & bioenergy 32: 829-837.

[21] Willebrand E, Ledin S, Verwijst T (1993) Willow Coppice Systems in Short Rotation Forestry: Effects of Plant Spacing, Rotation Length and Clonal Composition on Biomass Production. Biomass & bioenergy 4: 323-331.

[22] Bergkvist P, Ledin S (1998) Stem Biomass Yields at Different Planting Designs and Spacings in Willow Coppice Systems. Biomass & bioenergy 14: 149–156.

[23] Verwijst T (1991) Shoot Mortality and Dynamics of Live and Dead Biomass in a Stand of *Salix viminalis*. Biomass & bioenergy 1: 35-39.

[24] Willebrand E, Verwijst T (1993) Population Dynamics of Willow Coppice Systems and their Implications for Management of Short rotation Forests. The forestry chronicle 69: 699-704.

[25] Verwijst T (1996a) Stool Mortality and Development of a Competitive Hierarchy in a *Salix viminalis* Coppice System. Biomass & bioenergy 10: 245-250.

[26] Verwijst T (1996b) Cyclic and Progressive Changes in Short Rotation Willow Coppice Systems. Biomass & bioenergy 11:161-165.

[27] Nordh N-E (2005) Long Term Changes in Stand Structure and Biomass Production in Short Rotation Willow Coppice. Ph.D. Thesis. Acta Universitatis Agriculturae Sueciae 120. Swedish University of Agricultural Sciences, Uppsala, Sweden. 26 p.

[28] Tahvanainen L, Rytkonen VM (1999) Biomass Production of *Salix viminalis* in Southern Finland and the Effect of Soil Properties and Climate Conditions on its Production and Survival. Biomass & bioenergy 16: 103-117.

[29] Stolarski M, Szczukowski S, Tworkowski J, Bieniek A (2009) Productivity of Willow Coppice *Salix* spp. under Contrasting Soil Conditions. Electronic journal of polish agricultural universities 12: 1-10.

[30] Alriksson B (1997) Influence of site factors on *Salix* growth with emphasis on nitrogen response under different soil conditions. Ph.D. Thesis. Acta Universitatis Agriculturae Sueciae Silvestria 46. Swedish University of Agricultural Sciences, Uppsala, Sweden. 28p.

[31] Schaff SD, Pezeshki SR, Shields FD (2003) Effects of Soil Conditions on Survival and Growth of Black Willow Cuttings. Environmental management 31: 748-763.

[32] Verwijst T, Nordh N-E, Lundkvist A (2010) Effekter av sticklingsparametrar på grobarhet och tillväxt i salix. Grödor från åker till energiproduktion. Rapport 1144. Värmeforsk Service AB (Thermal Engineering Research Association), Stockholm,

Sweden. 47 p. Available: http://www.varmeforsk.se/rapporter?action=show&id=2522. Accessed 2012 April 18.

[33] Larsson S (1998) Genetic Improvement of Willow for Short Rotation Coppice. Biomass & bioenergy 15: 23-26.

[34] Karp A, Hanley SJ, Trybush SO, Macalpine W, Pei M, Shield I (2011) Genetic Improvement of Willow for Bioenergy and Biofuels. Journal of integrated plant biology 53: 151-165.

[35] McCracken AR, Dawson WM (1996) Interaction of Willow (*Salix*) Clones grown in Polyclonal Stands in Short Rotation Coppice. Biomass & bioenergy 10: 307-311.

[36] McCracken AR, Dawson WM (2003) Rust Disease (*Melampsora epitea*) of Willow (*Salix* spp.) Grown as Short Rotation Coppice (SRC) in Inter- and Intra-species Mixtures. Annals of applied biology 143: 381 393.

[37] Verwijst T (1993) Influence of the Pathogen *Melampsora epitea* on Interspecific Competition in a Mixture of *Salix viminalis* Clones. Journal of vegetation science 4: 717-722.

[38] Merila F, Heinson K, Kull O, Kuppel A (2006) Above-ground Production of Two Willow Species in Relation to Radiation Interception and Light Use Efficiency. Proceedings of Estonian Academy of Sciences. Biology, Ecology 55: 341–54.

[39] Caslin B, Finnan J, McCracken A (2010) Short Rotation Coppice Willow - Best Practice Guidelines. Teagasc, Crops Research Centre, Oak Park, Carlow, Ireland. 72 p. Available: http://www.teagasc.ie/forestry/docs/grants/WillowBestPracticeManual_2012.pdf. Accessed: 2012 April 18.

[40] Lindroth A, Cienciala E (1996) Water Use Efficiency of Short rotation *Salix viminalis* at Leaf, Tree and Stand Scales. Tree Physiology 16: 257-262.

[41] Verwijst T, Elowson S, Li X, Leng G (1996) Production Losses due to a Summer Frost in *Salix viminalis* Short Rotation Forest in Southern Sweden. Scandinavian journal of forest research 11: 104-110.

[42] Hollsten R, Arkelöv O, Ingelman G (2012) Handbok för Salixodlare. Jordbruksverket (Swedish Board of Agriculture), Jönköping, Sweden. 24 p. Available: http://www2.jordbruksverket.se/webdav/files/SJV/trycksaker/Pdf_ovrigt/ovr250.pdf. Accessed 2012 April 18.

[43] Börjesson P (1996). Energy Analysis of Biomass Production and Transportation. Biomass & bioenergy 11: 305-318.

[44] Weih M, Karacic A, Munkert H, Verwijst T, Diekmann M (2003) Influence of Young Poplar Stands on Floristic Diversity in Agricultural Landscapes. Basic and applied ecology 4: 149-156.

[45] Rosenqvist H (1997) Salixodling - Kalkylmetoder och Lönsamhet. Dissertation. Acta Universitatis Agriculturae Sueciae Silvestria 24. Swedish University of Agricultural Sciences, Uppsala, Sweden. 241p.

[46] Gustafsson J, Larsson S, Nordh NE (2009) Manual for SRC Willow Growers. Lantmännen Agroenergi, Sweden. 18 p. Available: http://www.agroenergi.se/PageFiles/135/ENG_Odlarmanual%20Salix%20(Reviderin g).pdf?epslanguage=sv. Accessed 2012 April 18.

[47] Danfors B, Ledin S, Rosenqvist H (1997) Short Rotation Willow Coppice - Growers Manual. Swedish Institute of Agricultural Engineering. JTI-informerar No. 1. 40 p.

[48] Volk T, Ballard B, Robinson DJ, Abrahamson LP (2004b) Effect of Cutting Storage Conditions during Planting Operations on the Survival and Biomass Production of Four Willow (*Salix* L.) Clones. New forest 28: 63-78.

[49] Burgess D, Hendrickson Q, Roy L (1990) The Importance of Initial Cutting Size for Improving the Growth Performance of *Salix alba* L. Scandinavian journal of forest research 5: 215-24.

[50] Shield IF, Macalpine W, Karp A (2008) The Effect of the Size of the Cuttings Planted on the Subsequent Performance of Three Contrasting Willow Cultivars for Short Rotation Coppice. Aspects of applied biology 90: 225-231.

[51] Verwijst T, Lundkvist A, Edelfeldt S, Forkman J, Nordh NE (2012) Effects of Clone and Cutting Traits on Shoot Emergence and Early Growth of Willow. Biomass & bioenergy 37: 257-264.

[52] Rossi P (1999) Length of Cuttings in Establishment and Production of Short Rotation Plantations of *Salix* 'Aquatica'. New forests 18: 161–77.

[53] Carpenter LT, Pezeshki SR, Shields FD (2008) Responses of Nonstructural Carbohydrates to Shoot Removal and Soil Moisture Treatments in *Salix nigra*. Trees 22: 737-748.

[54] Gage EA, Cooper DJ (2004) Controls on Willow Cutting Survival in a Montane Riparian Area. Journal of range management 57: 597-600.

[55] Edelfeldt S, Verwijst T, Lundkvist A, Forkman J (2012) Effects of Mechanical Planting on Establishment and Early Growth of Willow. Submitted to Biomass & bioenergy.

[56] Harper JL (1977) Population Biology of Plants. Academic Press: London and New York. 892 pp.

[57] Lundkvist A, Verwijst T (2011) Weed Biology and Weed Management in Organic Farming. In: Nokkoul R, editor. Research in Organic Farming. Rijeka: InTech. pp. 157-186. Available: http://www.intechopen.com/books/research-in-organic-farming/weed-biology-and-weed-management-in-organic-farming. Accessed: 2012 April 18.

[58] Abrahamson LP, Volk TA, Kopp RF, White EH, Ballard JL (2002) Willow Biomass Producer's Handbook. State University of New York College of Environmental Science and Forestry, Syracuse, NY, US. 31 p. Available: http://www.esf.edu/willow/pdf/2001%20finalhandbook.pdf. Accessed 2012 April 18.

[59] Verwijst T, Nordh N-E (2010) Effekter av skottnedklippning efter etableringsåret på produktionener första och andra omdrevet i salixodlingar. Grödor från åker till energiproduktion. Rapport 1136. Värmeforsk Service AB (Thermal Engineering Research Association), Stockholm, Sweden. 33 p. Available: http://www.varmeforsk.se/rapporter?action=show&id=2543. Accessed 2012 April 18.

[60] Labrecque M, Teodorescu TI, Babeux P, Cogliastro A, Daigle S (1994) Impact of Herbaceous Competition and Drainage Conditions on the Early Productivity of Willows under Short rotation Intensive Culture. Canadian journal of forest research 24: 493-501.

[61] Clay D, Dixon F (1997) Effect of Ground-cover Vegetation on the Growth of Poplar and Willow Short rotation Coppice. Aspects of applied biology 49: 53-60.

[62] Sage R (1999) Weed Competition in Willow Coppice Crops: the Cause and Extent of Yield Losses. Weed research 39: 399-411.

[63] Volk T (2002) Alternative Methods of Site Preparations and Coppice Management during the Establishment of Short rotation Woody Crops. Ph.D. Thesis. State University of New York, College of Environmental Science and Forestry, Syracuse, NY. 284 p. Available: http://en.scientificcommons.org/31526556. Accessed 2012 April 18.

[64] Smart L (2012) Cropping Systems for Shrub Willow Bioenergy Crops. Willowpedia, Cornell University, College of Agricultural and Life Science, Geneva, NY, US. Available: http://willow.cals.cornell.edu/Research/Cropping%20Systems.html. Accessed 2012 April 18.

[65] Moukoumi J, Farrell RE, Van Rees KJC, Hynes RK, Bélanger N (2012). Intercropping *Caragana arborescens* with *Salix miyabeana* to Satisfy Nitrogen Demand and Maximize Growth. BioEnergy Research: 1 14.

[66] van der Weide RY, Bleeker PO, Achten VTJM, Lotz LAP, Fogelberg F, Melander B (2008) Innovation in Mechanical Weed Control in Crop Rows. Weed research 48: 215-224.

[67] Helby P, Börjesson P, Hansen AC, Roos A, Rosenqvist H, Takeuchi L (2004) Market Development Problems for Sustainable Bioenergy in Sweden. (The BIOMARK Project). Report no. 38. Environmental and Energy System Studies, Lund University, Sweden. 193 p. Available: http://www.energimyndigheten.se/Global/Forskning/Energisystemstudier/Biomark%20project%20Projekt%2012010-1.pdf . Accessed 2012 April 18.

[68] Weih M (2001) Evidence for Increased Sensitivity to Nutrient and Water Stress in a Fast growing Hybrid Willow Compared with a Natural Willow Clone. Tree Physiology 21: 1141-1148.

[69] Ledin S (ed.) (1994) Gödsling av Salixodlingar: Ramprogram Energiskog. Rapport 1994: 25. Närings- och teknikutvecklingsverket (NUTEK), Stockholm, Sweden. 51p.

[70] Aronsson P & Rosenqvist H (2011) Gödslingsrekommendationer för Salix 2011. Slutrapport. Sveriges lantbruksuniversitet, Uppsala, Sweden, Stiftelsen lantbruksforskning, Stockholm, Sweden. 30 p. Available: http://pub.epsilon.slu.se/8466/1/aronsson_p_111125.pdf . Accessed 2012 April 18.

[71] Dimitriou I, Aronsson P (2011) Wastewater and Sewage Sludge Application to Willows and Poplars grown in Lysimeters – Plant Response and Treatment Efficiency. Biomass & bioenergy 35: 161-170.

[72] Rosenqvist H, Ness B (2004) An Economic Analysis of Leached Purification through Willow Coppice Vegetation Filters. Bioresource technology 94: 321-329.

[73] Pei MH, Royle DJ, Hunter T (1996) Pathogenic Specialization in *Melampsora epitea* var. *epitea* on *Salix*. Plant pathology 45: 679-690.

[74] Newcombe G (1996) The Specificity of Fungal Pathogens of *Populus*. In: Stettler RF, Bradshaw HD, Heilman PE, Hinckly TM, editors. Biology of *Populus*. Ottawa: NRC Research Press. pp. 223-246.

[75] Ramstedt M (1999) Rust Disease on Willows – Virulence Variation and Resistance Breeding Strategies. Forest ecology and management 121: 101-111.

[76] Åhman I (2000) Breeding for Resistane to Leaf Beetles attacking Biomass Willow in Europe. In: Isebrands JG, Richardson J, editors. 21st Session of the International Poplar Commission (IPC-2000): Poplar and Willow Culture: Meeting the Needs of Society and the Environment. USDA Forest Service. General Technical Report NC-215. St. Paul, MN: U.S. Dept. of Agriculture, Forest Service, North Central Forest Experiment Station. 226 p. Available: http://ncrs.fs.fed.us/pubs/gtr/gtr_nc215.pdf. Accessed 2012 April 18.

[77] Glynn C, Larsson S (2000) Rapid Gall Midge Adaptation to a Resistant Willow Genotype. Agricultural and forest entomology 2: 115-121.

[78] Björkman C, Höglund S, Eklund K, Larsson S (2000) Effects of Leaf Beetle Damage on Stem Wood Production in Coppicing Willow. Agricultural and forest entomology 2:131-139.

[79] Åhman I, Larsson S (1994) Genetic Improvement of Willow (Salix) as a Source of Bioenergy. Norwegian journal of agriculture sciences 18: Supplement 47-56.

[80] Åhman I, Larsson S (1999) Resistensförädling i Salix för Energiproduktion. Växtskyddsnotiser 63: 17-19.

[81] Nordh N-E (2010) Effekter på överlevnad och tillväxt vid förlängd skördesäsong av salix. Grödor från åker till energiproduktion. Rapport 1147. Värmeforsk Service AB (Thermal Engineering Research Association), Stockholm, Sweden. 31 p. Available: http://www.varmeforsk.se/rapporter?action=show&id=2490. Accessed 2012 April 18.

[82] Ceulemans R, McDonald A, Pereira JS (1996) A Comparison Among Eucalypt, Poplar and Willow Characteristics with Particular Reference to a Coppice, Growth Modeling Approach. Biomass & bioenergy 11: 215-231.

[83] Christersson L (1987) Biomass Production by Irrigated and Fertilized Salix Clones. Biomass 12: 83-95

[84] Larsson S, Lindegaard K (2003) Full Scale Implementation of Short Rotation Willow Coppice, SRC, in Sweden. Örebro, Sweden: Agrobränsle AB.

[85] Lindroth A, Båth A (1999) Assessment of Regional Willow Coppice Yield in Sweden on Basis of Water Availability. Forest ecology and management 121: 57-65.

[86] Larsson S (2001) Commercial Varieties from the Swedish Willow Breeding Programme Aspects of applied biology 65: 193-198.

[87] Lindegaard K, Parfitt RI, Donaldson G, Hunter T, Dawson WM, Forbes EGA, Carter MM, Whinney CC, Whinney JE, Larsson S (2001) Comparative Trials of Elite Swedish and UK Biomass Willow Varieties. Aspects of applied biology 65: 183-198.

[88] Walle IV, Van Camp N, Van de Casteele L, Verheyen K, Lemeur R (2007) Short Rotation Forestry of Birch, Maple, Poplar and Willow in Flanders (Belgium) I - Biomass Production after 4 Years of Tree Growth. Biomass & bioenergy 31: 267-275.

[89] Aylott MJ, Casella E, Tubby I, Street NR, Smith P, Taylor G (2008) Yield and Spatial Supply of Bioenergy Poplar and Willow Short Rotation Coppice in the UK. New Phytologist 178: 358-370.

Chapter 9

Artemia, a New Source of Animal Protein Ingredient in Poultry Nutrition

A. Zarei

Additional information is available at the end of the chapter

1. Introduction

In the nutritional behavior of single stomach animals, the origin of protein is important and its quality varies between different sources and animal origin is better than plant origin [7]. From the standpoint of salmonella contamination and due to high microbial potential, however, some of these proteins such as meat and bone meal ought to be used with caution. If heat treatment is not enough during fish meal processing, thiaminase will remain in fish meal and will cause harmful effects .Some researches indicated that thiamin will be reduced by thiaminase under special storage condition of ingredient before feeding animal [6,7] . Also, severe heating during meat and bone meal and fish meal processing to be confident that poultry by-product is safe, however some amino acids will probably degenerate and their bioavailability will decrease [5,7, 8,9] .

Although, meat & bone and fish meal have been used in poultry feeding, exclusively, but artemia biomass is also one of the animal proteins with high nutritional value which can be used in aquaculture and animal nutrition [1,4] .

The aim of this study is a survey of biology and characteristics of artemia and possibility of its usage in poultry nutrition as a protein source.

2. What is artemia?

Artemia or brine shrimp belongs to the animal kingdom, phylum of *arthropoda* , subphylum of *crustacean*, class of *branchiopoda*, order of *anostraca*, family of *artemidae* and genus of *artemia*. Linnaeus (1758) and Leach (1819) called it "*Cancer salinus*" and "*Artemia salina*", respectively. The latter name is because of the effect of salinity on morphological growth and development of artemia. Two species of artemia in Iran are: *Artemia urmiana* and

Artemia parthenogenetica. The first is native of Urmia lake and the second was observed in 12 regions of artemia habitats in Iran.

Artemia spreads in tropical and sub tropical regions in saline environments of the world, and over 500 artemia regions are discovered around the globe. Nine species of artemia were recognized in these regions.

More than two million kilograms of dried cysts of artemia with 0.4mm diameter are transacted in world markets every year. It is used as an aquaculture feed for hatched naplius. Uniformity of cysts and embryos with diapose has made artemia a unique source of aquaculture feeds. Artemia cysts can spread by wind and migratory birds.

Artemia contains 40-60% crude protein (dry matter basis) [11] .

3. Morphology and ecology

Morphologically, artemia has fragmented body with leave and wide shaped appearance. It's body consists of three compartments; head, thorax and abdomen, with total length of 8-10 mm and 10-12 mm in male and female respectively (Figure 1). Width of body is 4mm in both sexes. The exoskeleton of artemia is extremely thin (0.3-1μ) and flexible that called" chitin" and it is connected to muscle from inner surface.

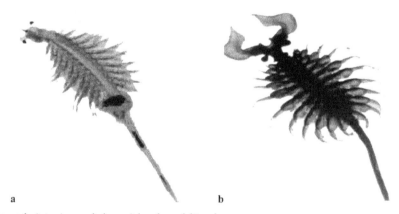

a   b

Figure 1. Artemia morphology a) female and b) male

The blood circulatory system of artemia is open.

This animal is euryhalin and can tolerate high concentration of salty habitat. There is a glandular organ in back of artemia neck that named; "salt gland" or "neck organ" .This organ exudates extra salt from the body to environment. Salty organ extinct at maturity and then this function is performed by exopodits of legs.

Although there are some limitations in the living environment of artemia, like high temperature and high salinity and drought, this animal can tolerate these conditions by producing cysts and going to diapose until the condition become suitable, and then it will continue its living.

Salinity and temperature are two important factors for growth and survival of artemia.

Artemia can tolerate salinity even more than 250 g/l and suitable range of temperature is 6-35°C.This crustacean has adapted itself with hard environmental conditions. In hypoxia, artemia increases the oxygen carrying capacity through the increase of the amount of hemoglobin. In this situation, body color turns red from original pale brown.

Physiological adaptation of artemia in high salinity is an effective defense method against predators using following mechanisms:

1. A powerful and effective osmoregulation system
2. Overcome high hypoxia in high salinity condition by higher pigmentation (inhalation pigments)
3. Production of embryos in diapose stage in cysts that can tolerate environmentally unfavorable condition.

4. Nutrition

From nutritional standpoint, artemia is non- selective filter feeder, and eats algae, bacteria, protozoa and yeasts, as long as feed particles diameter is not over 50-70 µm. Artemia feed can be alive or dead in artificial culture system. Artemia can uses bacteria and protozoa as feed sources, which grow in artemia culture medium. This protozoa (such as; *Candida, Rhodotorula*) can also be directly swallowed by artemia. The best algae's for artemia nutrition includes. *Dunaliella salina, Spirulina* and *Scenedesmus*. For artemia culture, agricultural products can be utilized such as; rice, corn, wheat, barley flours and their bran.

5. Reproduction

All bisexual species holds 42 chromosomes (2n = 42). *A.persimilis* holds 44 chromosomes (2n = 44) and *Artemia partenogenic* is diploid, triploid, tetraploid and even pantaploid. As a general rule, artemia populations are defined on the basis of the number of their chromosomes. However, contrary to mammalian, female artemia is heterogametic [3]. Artemia is produced by two ways: sexual and parthenogenesis (development of a new individual from an unfertilized egg). The mature female ovulates each 140 hours.

According to strain of artemia or method of living, it selects one of the following conditions; oviparous or ovoviviparous. In suitable situation of rising, reproduction trend is as larvae production (ovoviviparous) and in unsuitable situation of growth (salinity >50gm/lit and oxygen <5mg/lit) oviparous will occur. In the latter condition, growth of embryo will stop and enters diapauses. In suitable saline and nutritional conditions, females can produce 75 naplius each day and over its lift cycle (50 days), it reproduces 10-11 times.

In extreme hypoxia, due to increasing hemoglobin production, the color of artemia will change from light brown to yellow and then red.

Artemia cysts are spread by wind and birds. Earth pond or region of high salty water is suitable for culture and reproduction of artemia.

6. Different kinds of artemia from the nutritional point of view

From the nutritional point of view ,different kinds of artemia are:

- Decapsulated cysts
- Newly hatched nauplii
- Metanauplii and juvenile and adult
- Frozen and freeze – dried artemia

These forms of artemia are commonly used for newly hatched shrimps, sturgeons, trout, aquarium fishes and some crustaceans. Artemia biomass (consist of cysts and different living stages of artemia) is a suitable protein resource for other animals like poultry that consists of different stages of artemia growth.

7. Methods of artemia harvesting

According to artemia habitat, different biomass harvesting is utilized. In breeding pools and lake beaches, artemia is collected using a lace net that is fastened to two large floaters from each side (figure 2).

Because of phototropism characteristic, artemia can be collected easily by light source at night.

After harvesting, artemia biomass can be dried and cured under sunlight (Figure 3). Then the dried artemia will be milled before using in poultry diet.

Figure 2. Artemia biomass harvest

Figure 3. Flaked artemia

8. Chemical composition of artemia meal

The chemical composition of different kinds of artemia meal (dried at 50-60°C as sun cured or oven dried) is shown in Table 1. As shown in table 1, the chemical composition of 3 kinds of artemia meal (collected from different regions of Iran) is not identical. The quality of those ,depends on region, species, time of harvest and percentage of artemia mixture (artemia in different stages of living shows different compositions). So prior to using this ingredient, it must be analyzed for main nutrients.

Chemical composition		Kind of Artemia meal		
		ULAM	EPAM	GSLAM
Dry matter	g/kg	928	924	938
Crude Protein	g/kg	401.9	390.8	423.5
Gross Energy	MJ/kg	16.86	16.32	14.98
Crude Fat	g/kg	136	85.5	206.5
Crude Fiber	g/kg	36	18	28
Crude Ash	g/kg	240	287	284
Calcium (Ca)	g/kg	23.4	20.2	26.1
Phosphorus (P)	g/kg	11.1	8.6	14.2
Sodium (Na)	g/kg	12.1	9.6	16.4
Magnesium (Mg)	g/kg	3.3	4.1	3.1
Potassium (K)	g/kg	16.5	20.9	13.9
Iron (Fe)	mg/kg	1147.25	1642.75	437.75
Manganese (Mn)	mg/kg	53.78	132.45	84.08
Copper (Cu)	mg/kg	3.5	3.55	5.05
Zinc (Zn)	mg/kg	52.75	46.75	59

1- Zarei,A (2006) ,2- Urmia Lake Artemia Meal   , 3- Earth Pond Artemia Meal , 4- Ghom Salt Lake Artemia Meal

Table 1. Chemical composition of three kinds of artemia meal (ULAM[2], EPAM[3], GSLAM[4]) (as g/kg , MJ /kg or mg/kg – DM basis)[1]

9. Metabolizable energy of artemia meal

An experiment was designed to determine different classes of metabolizable energy (AME, AMEn, TME, TMEn) in artemia meals [13] .For determination of metabolizable energy of artemia meal and comparison with fish meal, samples gathered from 3 regions include : Urmia Lake Artemia Meal(ULAM), Earth Ponds Artemia Meal (beside urmia lake)(EPAM) and Ghom Salt Lake Artemia Meal(GSLAM). Then samples dried, milled and used in a biological experiment with fish meal. 20 Rhode Island Red cockerels with approximately same live weight used in Sibbald assay with completely randomized design with 5 treatments and 4 repetitions for determination of AME, AMEn, TME and TMEn.

Results showed there were significant differences between treatments from standpoints of metabolizable energy (P<0.05). ULAM and FM had highest ME and EPAM and GLAM had lowest ME. The highest TME belong to FM and the lowest TME pertained to EPAM. Except to EPAM that had the lowest TMEn, other treatments didn't have any differences between them.

10. Protein and amino acids digestibility of artemia meal

Result from *in vitro* and *in vivo* experiments showed that this ingredient has high quality of protein and the amount of digestibility was more than 90% [12] .

In order to determination of artemia meal's amino acid digestibility, five-week old male broiler chicks were given a semi-purified diet in which artemia meal was the sole source of protein. Apparent amino acid digestibility values of the assay diet, using ileum and excreta contents, were calculated using chromic oxide as indigestible marker. True digestibility values were calculated using endogenous output determined by feeding a nitrogen-free diet. The results showed (Table 2) that in determination of apparent amino acid digestibility of excreta, serine had the lowest (0.80) and methionine had the highest (0.92) digestibility, while glycine had the lowest (0.88) and arginine and leucine had the highest (0.95) apparent ileal digestibility. In measuring true excreta and ileal amino acid digestibility, alanine and glycine had the lowest (0.90 and 0.93) and methionine had the highest (0.96 and 0.99) digestibility, respectively. In general, the site of measurement had no effect on apparent or true amino acid digestibility of artemia meal [2] .

11. Artemia meal in broiler diets

In another experiment, different levels of protein from two kinds of artemia meal include artemia meal from Urmia lake and artemia meal from earth ponds beside Urmia lake with levels of 0, 25, 50, 75, 100 percent replaced to prue fish meal protein [12]. The experimental design was completely randomized with factorial method, include 10 treatments and 3 repetitions that in each repetition there were 10 one day-old male broilers from Ross 308 strain. This experiment was performed in 7 weeks and during and end of it, traits that related to broiler performance and carcass, was measured and analyzed. Results showed that effect of kind of artemia meal and effect of level of protein replacement weren't significant for feed intake. But interaction between these two was significant for this trait (P<0.05). The highest feed intake belong to Urmia lake artemia meal treatment with 50% level of replacement and the lowest feed intake related to treatment of without artemia meal (contain 5% fish meal). For body weight gain and feed conversion ratio, effect of kind of artemia meal and effect of level of protein replacement and effect of interaction between these two weren't significant. These effects weren't significant for all carcass traits and gastro intestinal parts exception for femur percent that treatment of without artemia meal (contain 5% fish meal) had a lowest percent to comparison with other treatments for this trait.

Amino acids	Apparent digestibility				True digestibility			
	Excreta	Ileal	SEM[1]	P[2]	Excreta	Ileal	SEM	P
Methionine	0.92	0.94	0.004	NS	0.96	0.99	0.004	0.09
Lysine	0.88	0.92	0.007	NS	0.92	0.96	0.007	NS
Threonine	0.85	0.90	0.013	NS	0.93	0.98	0.011	NS
Tryptophan	0.88	0.94	0.014	NS	0.90	0.97	0.017	NS
Arginine	0.89	0.95	0.008	0.09	0.93	0.98	0.008	NS
Isoleucine	0.88	0.94	0.011	NS	0.92	0.98	0.011	NS
Leucine	0.89	0.95	0.009	0.06	0.94	0.98	0.009	NS
Valine	0.87	0.93	0.011	NS	0.93	0.98	0.010	NS
Histidine	0.89	0.93	0.007	NS	0.95	0.97	0.007	NS
Phenylalanine	0.87	0.94	0.009	0.09	0.92	0.97	0.009	NS
Glycine	0.81	0.88	0.015	NS	-	0.93	-	-
Serine	0.80	0.89	0.018	NS	0.91	0.97	0.017	NS
Alanine	0.85	0.91	0.014	NS	0.90	0.94	0.014	NS
Aspartic acid	0.86	0.91	0.010	NS	0.91	0.94	0.005	0.09
Glutamic acid	0.87	0.93	0.014	NS	0.93	0.95	0.013	NS
Total	0.85	0.92	0.010	0.09	0.94	0.96	0.011	NS
CP(N×6.25)[3]	0.81	0.89	0.013	NS	0.89	0.94	0.012	NS

NS – Non Significant ; [1]- Standard Error of Mean ; [2] – Probability ; CP- Crude Protein ;N – Nitrogen [3] – The values (protein digestibility) were not corrected for uric acid.

Table 2. Apparent and true digestibility (coefficients) of artemia meal determined by sampling either excreta or ileum contents

12. Conclusion

Results of this studies revealed that artemia meal can be used as a feedstuff in poultry and other farm animal's diets because it has high level of protein and high protein digestibility. Compared with other animal proteins, artemia does not contain any feather, bone, hair or gastrointestinal tract components. In addition, in artemia production there is no requirement for high pressure and high temperature treatments which can influence protein quality. Artificial culture of artemia is easy and is possible everywhere.

Author details

A. Zarei*

Department of Animal Science, College of Agriculture and Natural Resources, Islamic Azad University- Karaj Branch, Karaj, Iran

13. References

[1] Abatzopoulos, TH.J, J.A. Beardmore., J.s.Clegg. and P.Sorgeloos. 2002. Artemia: Basic and Applied Biology. Kluwer Academic Publishers.

[2] Aghakhanian, P., A. Zarei, H. Lotfollahian and N. Eila. 2009. Apparent and true amino acid digestibility of artemia meal in broiler chicks.south African Journal of Animal Science.39 (1).

[3] Bowen, S.T.1964: The genetics of *Artemia salina*. IV. Hybridization of wild population with mutant stocks. Biol. Bull., 126, 333.

[4] Gilbert, V.S.1995.Introduction, biology and ecology of artemia. laboratory of aquaculture and artemia reference center university of Gent, Belgium.

[5] Johnson, M.L. and C.M. Parsons., G.C. Fahey, Jr., N.R. Merchen., C.G. Aldrich. 1998. Effects of species raw material source, ash content, and processing temperature on amino acid digestibility of animal by product meals by cecectomized roosters and ileally cannulated dogs. Journal of Animal Science.76:1112-1122.

[6] Klasing, K.C. 1998.Comparative Avian Nutrition.CAB International.

[7] Leeson. S. J.D.Summers. 2001. Nutrition of the Chicken.4 th edition. university Books.

[8] Parsons, C.M.1999. Protein quality and amino acid digestibility. Multi-state. Poultry meeting, May 25-27.

[9] Shirley, R.B. and C.M. Parsons. 2000. Effect of pressure processing on amino acid digestibility of meat and bone meal for poultry. Poultry Science.79:1775-1581.

[10] Sibbald, I.R., 1976.The true metabolizable energy values of several feedingstuffs measured with roosters, laying hens, turkeys and broiler hens. Poultry Science. 55:1459 -1463.

[11] Sorgeloos, P.1989. Two strains of Artemia in Urmia lake (Iran), Artemia Newsl., 13, 5.

[12] Zarei, A. (1).2006. Use of artemia meal as a protein supplement in broiler diet. International Journal of Poultry Science .5(2).142-148.

[13] Zarei, A. (2). 2006. Apparent and true metabolizable energy of artemia meal. International Journal of Poultry Science .5(7):627-628.

[14] Zarei, A. 2007. Determination of protein digestibility of animal protein feeds under condition of *in vitro* and *in vivo*, BSAS Annual Conference, 2-4 April 2007, Southport, England (oral presentation).

---

* Corresponding Author

Chapter 10

Characeae Biomass: Is the Subject Exhausted?

Carlos E. de M. Bicudo and Norma C. Bueno

Additional information is available at the end of the chapter

1. Introduction

Popularly known as stoneworts, brittleworts, muskgrass, muckworts or bass-weeds, Characeae are among the largest and most complex green algae. All common names come from some characteristics these plants may exhibit, such as the brittle, limestone (calcium carbonate) exoskeleton that can form on the external surfaces of the plant (e.g. *Chara vulgaris* and *Chara globularis*) and, particularly, from the distinctive smell of stale garlic emitted by the plant when crushed. It is important to note, however, that most charophyte species do not accumulate lime in observable amounts. The widespread misinterpretation that *Chara* plants generally form lime is understandable, since the two very common species above thrive in shallow waters where they are readily collected, forming spectacular extensive growths, and may solidify directly into a mail layer or onto a curious tufa rock, a porous limestone formed by deposits from springs. Plants accumulating lime become gray or whitish and quite opaque, whereas the many species without evident lime are generally soft, nearly transparent, with a glassy brilliance and rich green color.

Charophytes forms a significant part of the submerged vegetation of both natural and artificial systems represented by lakes, ponds, ditches, streams, canals, bog-pools, concrete tanks, reservoirs and excavations such as gravel pits, and are found on all continents except Antarctica. They are common in the littoral region of oligotrophic to moderately eutrophic water bodies (Kufel & Kufel 2002), and some authors (e.g. Krause 1985) consider these macrophytes indicators of water quality. *Nitella* specimens predominate in mildly acid water as in igneous rock areas, whereas *Chara*'s predominate in hard waters, but this is not a rule. They are characteristic of a disturbed habitat where periodic drastic changes create less favorable conditions for the growth of other algal species. They are often the first plants to colonize newly dug or cleared ponds and ditches, and some species are characteristic of ephemeral water bodies which dry up completely in the summer. The fast maturing charophytes have an advantage over the slow-growing macrophytes in such habitats. Charophytes are usually at a competitive disadvantage in shallow, moderately productive habitats, but tend to dominate in deeper water at low light intensities, particularly where the water has a high pH value. They are more often found in mesotrophic and eutrophic, hard water, calcium rich and low in phosphate waters. Charophytes may grow in silt, mud, peat or sand and they often form a dense carpet, known as a charophyte meadow, which restricts colonization by other macrophytes. The more common charophyte species do not die down during the winter. They have been recorded growing down to 60 m deep in clear water, but usually prefer depths between 1 and 10 m. In tropical countries such as Brazil, charophytes grow best in shallow water bodies, mostly in small reservoirs built for cattle disedentation, where they form dense carpets at the littoral zone of the reservoirs, usually at 20-40 cm depths. They may often grow intermingled with other macrophytes, mainly with water lilies (*Nymphaea* spp.), whose floating leaves they use to cut down the light intensity.

In size, they are generally moderately large, average shoots varying from 15 to 30 cm in height, but they may range from 5 mm to 2 m at extremes. Specimens of *Chara hornemannii* collected from the Rodrigo de Freitas Pond, in the city of Rio de Janeiro, ranged between 1.9-2 m tall.

The charophyte 'plant' or thallus is erect, central axis or 'stem' is branched and differentiated into a regular succession of nodes and internodes (Figure 1). Each node bears a whorl of branches of limited growth (the 'leaves' or branchlets), but branches capable of unlimited growth may arise axillary to the leaves. The axis consists of a chain of alternating

Figure 1. *Chara braunii* specimen showing the central axes branched and differentiated into a regular succession of nodes and internode, and oospores (black little rice-like structures at the verticelate branchlets) (source: Gutza Wikipedia)

long and short cells, the single long cells forming the internodes and the short, discoid cells forming the nodes. The single axial intermodal cell is commonly 1-4 cm long, but they may reach 50-60 cm in *Nitella cernua* and *Nitella translucens*. The intermodal cell is commonly 0.1-0.3 mm broad, but in the last two species, it may reach 2-3 mm broad. The plant is anchored by non-pigmented, single-celled processes, with or without a differentiation into nodes and internodes, the rhizoids, which penetrate the soil or substrate.

The importance of charophytes is indirect, as food for migratory waterfowl, protection of fish fry, and as a nuisance in shallow waters of reservoirs and recreational areas. They may also be used for sulfur baths, cattle food, fertilizers, scouring and filtering agents, and even for supposed control of mosquito larvae.

2. Methods for biomass estimation

Papers dealing with charophytes biomass are not numerous worldwide, and methods to measure that attribute are more or less standard.

Two boat-based and one in-water sampling method were used by Rodusky et al. (2005) to collect submersed aquatic macrophytes (SAV) as part of a long term monitoring program in Lake Okeechobee, Florida, U.S.A. The boat-based methods consisted of a ponar dredge used only to collect *Chara*, and an oyster tongs-like apparatus to collect all other SAV. The in-water method involved use of a 0.5 m² PVC quadrat frame deployed by a diver. Comparison of the three methods above showed no consistent pattern to the significant differences found in sampling precision between the three sampling methods, regardless of the geographical location, sediment type, SAV species or density.

To estimate charophytes biomass, the quadrat method is the most used one. According to the method, first a quadrate shall be delimited in the field, e.g. a 25 cm² (Westlake 1965, 1971; Krebs 1989). Within this quadrat, a 5 cm diameter (area 19.7 cm²) and 50 cm tall PVC tube is inserted. Tube wall must be perforated throughout the first basal 25 cm to allow water circulation and the gathering of the plants.

Once collected, material must be stored in glass vials (e.g. 50 ml volume) and taken to the laboratory. In the laboratory, charophytes must be gently washed and if necessary scrapped with a very soft brush to remove other algal material and sediments adhered to the plants. After washed and/or scrapped, the excess water must be dried with some paper towel and finally placed in a porcelain melter.

For the analytic procedure, the charophyte material must be calcinated at 550°C during 1 hour, then cooled in a desiccator and weighted using an analytical scale to have P_0. Immediately after, plants must be taken to an aerated oven at 65-70°C until no further weight change is observed for quantification of its dry weight (P_1), and after 1 hour calcination at 550°C for determination of its ash dry weight (P_2) (Hunter 1976). Determination of the ash free dry mass (AFDM) (P_3) is done using the mathematics $P_3 = (P_1 - P_0) - (P_2 - P_0)$. If total phosphorus (TP) is required, Strickland & Parsons (1965) method is to be used, i.e. the calcinated material is washed with 25 ml of HCl 1N, crushed and heated in a water-bath for 1 hour. After cooling, samples are diluted with 50-250 ml deionized water depending on the amount of calcinated material.

Palmer & Reid (2010) proposed a method they called 'invention' for the production of macroalgae to provide a sustained, economical source of biomass that may be used in various end-uses processes, including energy production. Their method provides specific combinations of macroalgae types, saltwater growth media compositions, and open pond water containers that resulted in biomass production beyond what may occur naturally without the required manipulation. Specifically, macroalgae that produce an exosqueleton in the presence of brackish water (e.g. stoneworths) have been found to provide excellent biomass production of at least 10 metric tons and up to 200 metric tons per acre per year under their method conditions.

Total phosphorus concentration is determined using a spectrophotometer. Another possibility for TP determination is by the molybdenum blue colorimetric method (Murphy & Riley 1962) after digestion with $K_2S_2O_8$ in an autoclave at 120°C for 30 minutes (APHA 1995). Total nitrogen (TN) can be determined using spectrophotometry, based on the Koeofell colorimetric method. Calcium and magnesium can also be determined using a spectrophotometer, however, based on the Calmagite colorimetric method.

3. Charophytes biomass

The vast majority of papers published on charophytes worldwide, deals with their taxonomy and systematics. Comparatively, very few papers were published dealing with their biology, including citology, genetics, ecology and physiology.

Measure of biomass is one possibility to estimate the macrophyte's capacity to photosynthetize (Wetzel 1964) and the most used. Other possibilities include population density and biovolume. According to Wetzel (2001), the submersed macrophytes biomass is low if compared to that of other plants. The importance of the charophytes living at the littoral zone of lakes is directly related to the amount of submersed biomass, spatial structure and these plants association with other submersed and emerged macrophytes.

Literature regarding charophytes biomass is not rare neither profuse worldwide and dates mostly from 1980 on, when eutrophication was recognized to be one of the most important events of the century. While not profuse, literature available consents a pretty good overview on the subject.

3.1. Seasonal variation

In the temperate region of the World, aquatic macrophytes show very sharp annual variation, with a growth season of their aerial biomass during the spring and summer, and another season of the underground biomass and detritus accumulation during the fall and winter (Esteves & Camargo 1986). In the tropical region, however, deterministic of the aquatic macrophytes biomass seasonality are the rainy and dry periods (Esteves 2011). Very little was done, however, up to now regarding the charophytes biomass seasonality in the tropics. Perhaps the only contribution in this regard is the work by Carneiro et al. (1994), who studied the extensive *Chara hornemanii* beds prospering in the Piratininga Lagoon, State of Rio de Janeiro, southeast Brazil at the depth from 0.30 cm to 1 m, and realized that N and P inputs, low water turnover and low water column depth favored growth of phytoplankton, macroalgae and aquatic macrophytes, including charophytes. The same authors also observed a very clear seasonal behavior of the charophyte population that stared during the winter and lasted until the beginning of summer, when the alga covered about 60% of the lagoon sediments. During the summer, the alga biomass reached 500 mg m^{-2} (Carneiro et al. 1994).

Using aerial photographs and field work in brackish water lagoons of Åland Island, Finland, Berglund et al. (2003) observed seasonal and interannual growth, distribution and biomass variation of some charophyte species. According to the last authors, filamentous green algae contributed with 45-70% of the total biomass studied, charophytes with 25-40% and vascular plants with 3-18%. The biomass peak was reached in July and August, and the average biomass was negatively correlated with the charophytes exposition to direct sun light, i.e. the charophyte coverage was greater when their exposition to solar radiation was low, being highly affected by the presence of filamentous algae.

Seasonal changes in the biomass of a monospecific community (*Chara globularis*) and of several communities with high charophyte coverage (*Chara globularis–Myriophyllum alterniflorum*, *Chara globularis–Potamogeton gramineus* and *Nitella translucens–Potamogeton natans*) were studied monthly, from May 1996 to June 1997, by Fernández-Aláez et al. (2002) in three shallow lakes in northwest Spain. Weather and hydrological regime strongly influenced the seasonal biomass patterns and the between-the-year differences in the biomass of the macrophytes. The *Chara globularis* community biomass showed a bimodal pattern, with maximum in mid-July (128 g DW m^{-2}) and late autumn (165 g DW m^{-2}). *Chara globularis* overwintered as a green plant and during the subsequent growth period characterized by high temperature and low rainfall reached a maximum of 305 g DW m^{-2} in June 1997. The highest biomass of *Chara globularis* in the *Chara–Myriophyllum* community was reached in July (Lake Sentiz 160 g m^{-2}, Lake Redos 204 g m^{-2}), while the minimum (Lake Sentiz 10 g m^{-2}; Lake Redos 3 g m^{-2}) was recorded in February or March. *Myriophyllum alterniflorum* (average biomass 95 g m^{-2}) was a better competitor than *Chara globularis* in Redos lake and appeared to be favored by the early beginning of the growing season in 1997 and by the later increase in the water level. *Nitella translucens* biomass (average 64 g m^{-2}) showed a high stability during the entire study period, but lacked a well-defined seasonal pattern. *Potamogeton natans* had a marked maximum biomass in August (426 g m^{-2}). Although the stability of the *Potamogeton natans* population was low, shading did not have a significant influence on the development of *Nitella translucens* biomass.

Torn et al. (2006) measured the seasonal dynamics of the biomass, elongation growth and primary production rate of *Chara tomentosa* in Rame Bay, NE Baltic Sea, a shallow and semi-enclosed sea inlet on the western coast of the Estonian mainland, during the vegetation period of 2002. Their measurements showed extremely high plant heights (up to 1.42 m) and biomass values (5.2 kg (w.w.) m^{-2}) indicating the importance of the charophyte for the aquatic ecosystem. Torn et al. (2006) observed that the apical part of the plants grew more intensively from early spring to midsummer, whereas that of the subapical one was very low during the entire study period. The plant's net primary production rate peaked in July (43.4 mgO g(d.w.)$^{-1}$ 24 h^{-1}), remarkably lower rates being measured in May and September. The elongation growth and primary production were not correlated with the water nutrient concentrations and temperature. As the active growth of *Chara tomentosa* takes place during a relative short period at the beginning of summer, the amount of available solar radiation and the temperature levels during this sensitive time may have had a significant effect on the community in the same year (Figure 2).

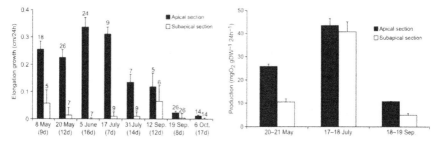

Figure 2. Left: seasonal variation in average elongation growth + S.E. of *Chara tomentosa*. Date of measurement and duration of the experimental period are indicated in between parenthesis. Number of replicates is indicated at the top of each bar. **Right**: Seasonal variation in average diurnal net primary production rate + S.E. of *Chara tomentosa*. In each period photosynthesis was measured in the period of 24 h replicated 3-fold (Torn et al. 2006).

Seasonal growth of *Chara globularis* var. *virgata* caused a regular summer depletion of Ca^{2+} and HCO$_3^-$ by associated CaCO$_3$ deposition, and a more extreme and unusual depletion of K$^+$ was followed over three years (1985-1987) by Talling & Parker (2002) in a shallow upland lake (Malham Tarn) in northern England. Chemical analysis of the *Chara globularis* var. *virgata* biomass and of the underlying sediments indicated a large benthic nutrient stock that far surpassed that represented by the phytoplankton. Growth in the *Chara globularis* var. *virgata* biomass and the magnitude of the water-borne inputs influenced removals of Ca^{2+}, K$^+$ and inorganic N. According to Talling & Parker (2002), several features of Malham Tarn are suggestive in relation to the general case of phytoplankton-phytobenthos interaction and possible long-term change. So, the low P concentrations in the open water are probably linked to the fairly low phytoplankton abundance and influenced by the dense benthic *Chara globularis* var. *virgata* with a major capacity for P uptake. Also, the additional *Chara globularis* var. *virgata* capacity for K$^+$ uptake led to a major seasonal reduction of concentration in the lake water and outflow, of a magnitude rarely if ever recorded elsewhere. The annual growth of *Chara globularis* var. *virgata* seemed to involve further translocation of N, P and K from stocks in the sediments.

3.2. Impact of climatic fluctuations on the biomass

Sender (2008) studied the long term changes of the macrophytes structure in the Lake Moszne located in the Poleski National Park in Poland. Lake Moszne is a relatively small (17.5 ha), distrophic and shallow (1 m) water body. The lake is not connected with the size, nor with the depth of the reservoir, thus depending on the climatic conditions as well as on the economic and recreational activities, and on the hydro-technical changes imposed to the lake (Sender 2008). As a result, a distinct decrease of the plant association variety was observed, as well as changes in their qualitative composition. In fact, changes in qualitative and quantitative structure of lake Moszne macrophytes were probably caused by both abiotic and biotic factors. The macrophytes structure was subject to fluctuation, the changes indicating notable growth of water trophy. The biomass of macrophytes also showed an increase tendency. Nowadays, the structure of vegetation of the lake does not show the typical features for distrophic lakes.

It is well known that algal populations are often present in considerable and varying densities within shallow lakes, as both planktonic and benthic components (Talling & Parker 2002), and that shallow lakes have become the archetypical example of ecosystems with alternative stable states (Scheffer & van Nes 2007). Moreover, that shallow lakes may switch from a state dominated by submersed macrophytes to a phytoplankton-dominated state when a critical nutrient is exceeded (Kosten et al. 2011). Last authors explored how climate change affected that critical nutrient concentration by linking a graphical model to data from 83 lakes along a large climate gradient in South America. Their data indicated that in warmer climates, submersed macrophytes may tolerate more underwater shade than in cooler lakes, although the relationship between phytoplankton biomass and nutrient concentrations did not change consistently along the climate gradient. According to Kosten et al. (2011), in several lakes in the warm and intermediate regions, submersed macrophytes were found until relatively greater depths than in the cool regions, taking the available light at the sediments surface into account.

Rip et al. (2007) is an excellent case-study of how temporal pattern of precipitation and flow from land to water, may give a coherent, quantitative explanation of the observed dynamics in P, phytoplankton, turbidity and charophytes. Studying the external P load to a wetland with two shallow lakes in the Botshol Nature Reserve, The Netherlands the above authors observed that P load reduction resulted in a rapid decrease of phytoplankton biomass and turbidity, and after four years in an explosive charophyte growth. Such a clear water state, however, was unstable and the ecosystem alternated between clear, high-vegetation and turbid, low-vegetation states. Rip et al. (2007) used a water quality processes' model in conjunction with a 14-year nutrient budget for Botshol to determine if fluctuations in precipitation and nutrient load effectively caused the ecosystem instability. Their results indicated that during wet winters when groundwater level rose above surface water level, P from runoff was stored in the lake sediments and banks (Figure 3). Stored P was released the following spring and summer under anaerobic sediments conditions, thus resulting in an increase of phytoplankton density and light attenuation in the water column. Also, in years with high net precipitation, flow from land to surface water also transported humic acids, further increasing light attenuation. Conversely, in years with dry winters, P and humic acid loads to surface water were reduced, and growth of submersed macrophytes enhanced by clear water. Rip et al. (2007) concluded by stating that global warming caused winters in the Netherlands to become warmer and wetter during the last 50 years, consequently increasing flow from land to water of humic acids and P and, ultimately, enhancing instability of charophyte populations. Finally, in the first half of the 20th Century interannual variation in precipitation was not sufficient to cause large changes in the internal P flux in Botshol, and submersed macrophytes population were stable.

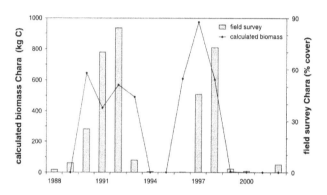

Figure 3. Calculated *Chara* biomass as model results and field surveys at subarea I of the Botshol Natural Reserve for 1989-2002 (Rip et al. 2007).

Recently, Salmaso et al. (2012) studied the combined effects of nutrient availability and temperature on phytoplankton in large and deep lakes of the Alps, lakes Garda, Iseo, Como, Lugano and Maggiore. A significant effect of temperature fluctuations and trophic status on the development of the main groups of cyanobacteria and eukaryotic phytoplankton was observed. However, high positive relationships of nutrient availability with temperature were found only in a few algal groups including charophytes, chlorophytes, dinophytes and, partly, cyanobacteria. Their results have implications in the evaluation of the impact of different climatic scenarios in lakes of different trophic status, suggesting a net increase of only selected eutrophic- or eurytrophic sensitive groups with increasing water temperature in more enriched systems.

3.3. Influence of depth and transparency

Once established, aquatic macrophytes have a positive effect on the transparency of water through several buffer mechanisms (Stephen et al. 1998). Furthermore, the presence of charophytes has been associated with the maintenance of clear water, and changes from a state of clear to turbid water have been associated with the eutrophication of the environment (e.g. Blindow et al. 1993, Kufel & Kufel 1997).

Steinman et al. (1997) studied the influence of water depth and transparency on the charophyte biomass distribution in the southern end of the subtropical Lake Okeechobee, U.S.A. Their first survey (August 1994) was conducted on 47 stations within the 3-Pole Bay. Subsequent surveys (November 1994-December 1996) were conducted on a monthly or bimonthly basis on 7 stations. According to the authors, the distribution and abundance of *Chara* population in the lake showed a marked seasonal phenology, although there were notable differences in biomass among the years and stations. *Chara* plants were observed only in August, September and October, and in 1996 also in November. Also, biomass never exceeded 20 g AFDM m^{-2} and declined significantly from 1994 to 1996. The charophyte biomass was inversely related to the water depth and positively related to the Secchi disc depth, suggesting that irradiance strongly influenced the charophyte distribution in the lake, a hypothesis that was confirmed by data they collected from photosynthetic measurements and photosynthesis-irradiance curves (Steinman et al. 1997).

The role of charophytes in increasing the water transparency was also studied by Nõges et al. (2003). Under the frame of the EC project ECOFRAME, last authors worked out the water quality criteria for two shallow lakes of the Vooremaa landscape protection area, Central Estonia. Lake Prossa is a macrophyte-dominated system with an area of 33 ha and a mean depth of 2.2 m. Most of its bottom is covered by a thick mat of charophytes all year round. Lake Kaiavere is located 10 km far from Lake Prossa, is much larger (250 ha, mean depth 2.8 m) and is phytoplankton-dominated. Nevertheless, the nutrient dynamics was very similar in the two lakes (Figure 4). The first vernal phytoplankton peak was expressed in reduced Secchi depth in both lakes. After that peak, the water became clear in Lake Prossa, but remained turbid in Lake Kaiavere. Towards fall, the individual mean weight of zooplankton decreased in Lake Prossa, the *Chara* lake, but remained smaller than in the plankton dominated one (Lake Kaiavere) (Figure 5). Therefore, zooplankton grazing would initiate the clear-water phase in the *Chara*-lake, but other factors were needed for its maintenance. Another factor that showed a clear difference between the two lakes was the carbonate alkalinity that was rather stable or even increased during the spring in the phytoplankton-dominated lake, while it decreased by nearly 50% between April and July in the *Chara*-lake. The reduced sediment resuspension and the possible allellopathic influence of charophytes on phytoplankton remain the main explanations for the maintenance of the extensive clear-water period in the *Chara*-lake.

Blindow & Schütte (2007) worked with material from fresh and brackish water in Sweden and found out that both turbidity and salinity acted as stress factors on *Chara aspera*. According to the last authors, in clearwater lakes the species can occur in high densities and reach deep water, where the ability to hibernate as a green plant together with shoot elongation may further extend the lower depth limit. In turbid lakes, the plants can still form dense mats, but are restricted to shallow water due to the poor light availability, although shoot elongation may allow a certain extension of the depth range (Figure 6).

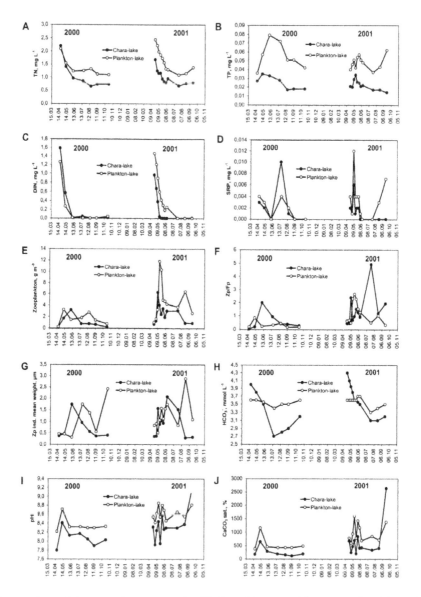

Figure 4. Seasonal dynamics of some chemical and biological features in lakes Prossa (*Chara*-lake) and Kaiavere (plankton-lake). A – total nitrogen, B – total phosphorus, C – dissolved organic nitrogen, D – soluble reactive phosphorus, E – zooplankton/phytoplankton biomass ratio, G – zooplankton mean individual weight, H – hydrocarbonate concentration, I – pH, and J – calcite saturation level (Nõges et al. 2003).

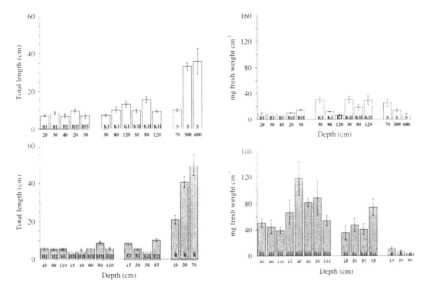

Figure 5. Left: total length of *Chara aspera* (mean values + S.E., n = 5), determined at all sites, depth ranges and sampling occasions. Above freshwater sites, bellow brackish water sites. K – Lake Krankesjön, B – Lake Börringesjön, S – Lake Storacksen, E – Edenryd bay, H – Höllviken bay, R – Redensee bay. I and II – first and second sampling occasions, respectively. Right: fresh weight:total length ratio of *Chara aspera* determined at all sites, depth ranges and sampling occasions (Blindow & Schütte 2007).

A field study conducted from July 2003 to May 2005 in the Myall Lake, a brackish shallow lake in New South Wales, Australia, revealed that *Chara fibrosa* var. *fibrosa* and *Nitella hyalina* occurred in areas of the entire lake that were deeper than 50 cm. Also, more fresh shoots were obtained during the winter (water temperature 13-16ºC), thus suggesting that winter may be their preferred growing season. Their biomass varied from 0 to 321 g DW m^{-2}, their maximum biomass being displayed between 1 and 2.5 m depth (Asaeda et al. 2007). These authors also observed that charophyte's shoots were longer in deeper waters, varying from c. 30 cm at 1 m depth to 60-90 cm between 2 and 4 m depth. Plants growing in shallow depths had shorter internodes implying a shorter life cycle of shoots. Also, nodal spacing was relatively regular in contrast to its deeper water counterparts although spacing tended to increase at locations farther from the apex (Figure 7). Finally, numbers of oospore and antheridia were higher in shallower water reaching their maximum at around 80 cm.

Chambers & Kalff (1985) used original data from eight lakes in southern Quebec, Canada and literature data from other lakes throughout the World to predict the maximum depth of charophytes colonization and the irradiance over the growing season at the maximum depth of colonization, concluding that the depth distribution of the aquatic macrophyte communities is quantitatively related to Secchi depth. According to regression models proposed in Chambers & Kalff (1985), natural distribution of aquatic macrophytes is restricted to depths of less than 12 m, whereas charophytes can colonize to great depths and up to a predicted 42 m in the very clearest lakes (Secchi depth 28 m).

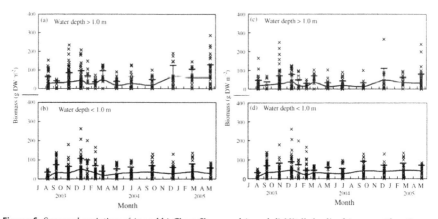

Figure 6. Seasonal variation of (a and b) *Chara fibrosa* and (c and d) *Nitella hyalina* biomass at location deeper than and shallower than 1 m. "X" markers denote individual measurements, the thick solid line represents monthly means, and short flat bars indicate standard deviations (mean + 1 S.D.) (Asaeda et al. 2007).

3.4. Nutrients

The concentrations of N, P and C in the above-ground biomass of 14 dominant macrophyte species (including *Chara globularis* and *Nitella translucens*) in seven shallow lakes of NW Spain were measured by Fernández-Aláez et al. (1999) that found significant differences for the three nutrients among the species and among the groups of macrophytes. The charophytes showed the lowest P (0.053% dry weight) and C (35.24% dry weight) content. Also, only the charophytes exhibited a strong association between N and P ($r = 0.734$, $p < 0.0001$), reflecting an important biochemical connection in these species.

Phosphorus was established as a limiting factor of all the macrophytes (N:P = 35:1), especially charophytes, in which it was below the critical minimum. Siong et al. (2006) used sequential P fractionation to study the nutrient speciation in three submersed macrophytes species, *Chara fibrosa*, *Najas marina* and *Vallisneria gigantea*, and the implications for P nutrient cycling in the Myall Lake, New South Wales, Australia. The mean TP of both *Najas marina* and *Vallisneria gigantea* was significantly higher than that of *Chara fibrosa*, even when the comparison made was based on the ash-free dry weight (AFDW). However, P co-precipitation with calcite ($CaCO_3$) induced during intense periods of photosynthesis occurs in hard water lakes, and this indirect mechanism of reducing P bioavailability in the water column may have been underestimated in assessing *Chara* beds acting as nutrient sink in shallow lakes. According to their results, besides the indirect mechanism above, P in the water column was also directly co-precipitated with encrusted calcite along the charophyte intermodal cell, and such a calcification should be regarded as a positive feedback in stabilizing *Chara* dominance in lakes. Siong & Asaeda (2009) studied the effect of Mg on the charophyte calcite encrustation, and assessed whether charophytes growing on the non-calcareous sediments of the Myall Lake could function as an effective nutrient sink for P in a similar manner to charophytes growing on the calcareous sediments of freshwater calcium-rich hard water systems. According to the last authors, calcification of *Chara fibrosa* was significantly inhibited by Mg in the water column and, consequently, reduced the formation of Ca-bound P that has a potential sink for P. However, a large percentage of non-bioavailable forms of P in the lake sediments suggested that P sink was through burial of dead organic matter and subsequent mineralization process.

The inorganic phosphorus concentration was not yet significantly related to the charophyte biomass. Palma-Silva et al. (2002) observed that the charophyte community (*Chara angolensis* and *Chara fibrosa*) sometimes occupied the entire benthic region in the Imboacica coastal lagoon in Brazil, and presented a large variation in C:N:P ratio. Results of their investigation (samples taken in March, April, May, July and October 1997) indicated that the charophytes fast growth may have absorbed a great amount of the nutrients entering the lagoon. Values of nutrient concentrations in the charophytes biomass were, according to those authors, within the expected range for the group, with the most eutrophic sampling station in the lake showing the highest N and P values. C:N:P ratios presented high values, and the biomass values were higher in the less eutrophic areas. The biomass reached maximum values of between 400 and 600 g DW m^{-2}, and the C:N:P ratio varied from 51:7:1 to 1603:87:1, indicating that the two *Chara* species may grow in a wide range of nutrient concentration. The same authors concluded that the charophyte community would be responsible by the nutrient decrease in the water column and keeping the water clear after drawdowns (Palma-Silva et al. 2002).

Several authors concluded that the nutrient kinetics favor the phytoplankton growth over *Chara*, thus assuming a P-limited condition. Therefore, although nutrient concentration may influence the charophyte phenology and abundance, light appeared to be a stronger regulator in the Okeechobee Lake. Schwarz & Hawes (1997) also observed the influence of the water transparency on the variation of the charophyte biomass in the Coleridge Lake, New Zealand. In the latter lake, total algal biomass did not surpass 180 g DW m^{-2} between 5 and 10 m depth. Pereyra-Ramos (1981) worked with seven charophyte species collected from Polish lakes and observed an increase of their fresh dry weight during the summer (July): *Chara rudis* 2.07 kg m^{-2}, *Chara vulgaris* 1.61 kg m^{-2}, *Chara contraria* 0.54 kg m^{-2}, *Chara fragilis* 0.39 kg m^{-2}, *Chara jubata* 0.37 kg m^{-2}, *Chara tomentosa* 0.28 kg m^{-2} and *Nitellopsis obtusa* 0.24 kg m^{-2}. Together, the charophytes represented 53% of the total submersed macrophytes biomass, 28% of *Elodea* sp. and 8% of *Ceratophyllum demersum*, two submersed macrophytes. According to Howard-Williams et al. (1995), *Chara corallina* biomass in deep (average 90 m depth) New Zealand lakes ranged around 300 g DW m^{-2}. Bakker et al. (2010) registered a strong decline of the *Chara* sp. biomass under the nutrient enriched condition of Lake Loenderveen, Norway. Similar situation was already detected by Blindow et al. (1993) and van de Bund & van Donk (2004) for other water bodies.

3.5. Trace contaminants

The Anthropocene period is characteristic by rapid urbanization, industrialization, mining activities, metal ore refining, agricultural chemicals, liquid and solid wastes, resulting in heavy metal pollution of water and land resources. There has been an increasing load of heavy metals (Cu, Zn, Cd, Cr, Hg and Ni) in the aquatic ecosystems, which in turn are being assimilated and transferred within food chains by the process of biomagnification. The problem with the heavy metals is their non-biodegradable nature. The conventional methods used to remove metal ions include chemical precipitation, lime coagulation, ion exchange, reverse osmosis solvent extraction, aeration, chemical oxidation, electrodialysis, ultra filtration, and chlorination (Rich & Cherry 1987).

Research was carried out recently to evaluate the metal accumulation in charophytes. Hence, Bibi et al. (2010) investigated the effects of Cd, Cr and Zn on the growth of *Nitella graciliformis* and their bioaccumulation in the plant under laboratory conditions. Charophyte specimens were exposed to different Cd, Cr and Zn concentrations, and it was observed that the heavy metals concentrations in the plant increased with the increasing metals concentrations in the mediums. As a result, negative growth occurred and the internode elongation was reduced when exposed to these metals at any concentration, however, intracellular *Nitella gracilliformis* has a potential for accumulating Cd, Cr and Zn. Bibi et al. (2010) concluded their investigation by stating that their study should be an integral part of the sustainable development of ecosystems and pollution assessment programs.

Absorption processes are being widely used for the removal of heavy metals from aqueous solutions. According to Shaikh Parveen & Bhosle Arjun (2011), use of various products has been widely investigated in the recent years as an alternative for the currently expensive methods of water treatment, and some natural products can be effectively used as a low cost absorbent. The above mentioned authors conducted batch studies of *Hydrilla* sp. and *Chara* sp. to evaluate the uptake of Cr from aqueous solutions. They found out that about 91.7% removal was obtained with 2 mg L^{-1} of *Chara* sp. at 2 mg L^{-1} Cr concentration after a period of seven days at pH 4. Their results also indicated that the metal removal increased as the days were extended, however, with the increasing contact time *Hydrilla* sp. proved to be better than *Chara* sp. in the Cr removal.

4. Final remarks

As it was mentioned before, literature on charophytes biomass is not rare neither profuse worldwide and dates mostly from 1980 on, when eutrophication was recognized to be one of the most important events of the century. Despite of not being profuse, literature available consents a pretty good overview on the subject.

In the temperate region of the World, aquatic macrophytes show very sharp annual variation, with a growth season of their aerial biomass during the spring and summer, and another season of the underground biomass and detritus accumulation during the fall and winter. Very little, however, was done up to now regarding the charophytes biomass seasonality in the tropics. The single paper published based on charophytes from the tropical region defined, however, deterministic of the aquatic macrophytes biomass seasonality the rainy and dry periods. Water temperature and rain precipitation are, nevertheless, somewhat connected to each other, since the rainy season in the tropics somewhat coincides with the high temperature season.

A climate gradient in South America was studied, indicating that in warmer climates, submersed macrophytes may tolerate more underwater shade than in cooler lakes. Moreover, in several lakes in the warm and intermediate regions, submersed macrophytes were met until relatively greater depths than in the cool regions, taking the available light at the sediments surface into account. According to a very detailed long term study, global warming has been causing winters in the Netherlands to become warmer and wetter during the last 50 years, consequently increasing flow of humic acids and P from land to water that, ultimately, has been enhancing instability of charophyte populations. Such studies conclusion is that in the first half of the 20th Century interannual variation in precipitation was not sufficient to cause large changes in the internal P flux, and submersed macrophytes population was stable.

The presence of charophytes has been associated with the maintenance of clear water, and changes from a state of clear to turbid water have been associated with the eutrophication of the environment. Original data from eight lakes in southern Quebec, Canada and literature data from other lakes throughout the World were used to predict the maximum depth of charophytes colonization and the irradiance over the growing season at the maximum colonization depth, concluding that the depth distribution of the aquatic macrophyte communities is quantitatively related to the Secchi depth. Regression models using the same information above, defined that natural distribution of aquatic macrophytes is restricted to depths of less than 12 m, whereas charophytes can colonize to great depths and up to a predicted 42 m in the very clearest lakes.

The inorganic phosphorus concentration was not yet significantly related to the charophyte biomass. Concentrations of N, P and C in the above-ground biomass of 14 dominant macrophyte species (*Chara globularis* and *Nitella translucens* included) in seven shallow lakes of NW Spain pointed to significant differences for the three nutrients among the species and among the macrophytes groups, the charophytes showing the lowest P and C content. Also, only the charophytes showed a strong association between N and P.

Only recently some research has been carried out to evaluate the metal accumulation in charophytes. Therefore, charophyte specimens were exposed in laboratory experiments to different Cd, Cr and Zn concentrations, showing that the heavy metals concentrations in the plant increased with the increasing metals concentrations in the cultivation mediums used. As a result, negative growth occurred and the internode elongation was reduced when exposed to these metals at any concentration, however, intracellular *Nitella gracilliformis* revealed a potential for accumulating Cd, Cr and Zn.

Summarizing, all research done up to now on the charophytes biomass is still very punctual, i.e. they most often focused one special environment under very specific conditions. There are very few studies focusing a larger time scale and comparing several localities. In the last cases, results are much more consistent. The scientific community needs much more studies, to be able to formulate generalizations. In other words, despite of producing some important information, study of charophytes biomass is far from being exhausted, on the contrary they have just started.

Author details

Carlos E. de M. Bicudo
Instituto de Botânica, São Paulo, SP, Brasil

Norma C. Bueno
Universidade Estadual do Oeste do Paraná, Cascavel, PR, Brasil

Acknowledgement

CEMB is very much indebted to CNPq, Conselho Nacional de Desenvolvimento Científico e Tecnológico for partial financial support (Grand n⁰ 309474/2010-8).

5. References

[1] APHA, American Public Health Association (1998) Standard methods for the examination of water and waste water. American Public Health Association, Washington, D.C. 20th edition.

[2] Asaeda T, Rajapakse L & Sanderson B (2007) Morphological and reproductive acclimatation to growth of two charophyte species in shallow and deeper water. Aquatic Botany 86: 393-401.

[3] Bakker ES, van Donk E, Declerck SAJ, Helmsing NR, Hidding B & Nolet BA (2010) Effect of macrophyte community composition and nutrient enrichment on plant biomass and algal blooms. Basic and Applied Ecology 11: 432-439.

[4] Berglund J, Mattila J, Rönnberg O, Keikkila J & Bondsdorff E (2003) Seasonal and inter-annual variation in occurrence and biomass of rooted macrophytes and drift algae in shallow bays. Estuarine Coastal Shelf Science 56: 1167-1175.

[5] Bibi MH, Asaeda T & Azim E (2010) Effects of Cd, Cr, and Zn on growth and metal accumulation in an aquatic macrophyte, *Nitella graciliformis*. Chemistry and Ecology 26(1): 49-56.

[6] Blindow I, Andersson G, Hargeby A & Johansson S (1993) Long-term pattern of alternative stable states in two shallow eutrophic lakes. Freshwater Biology 30: 159-167.

[7] Blindow I & Schütte M (2007) Elongation and mat formation of *Chara aspera* under different light and salinity conditions. Hydrobiologia 584: 69-76.

[8] Carneiro MER, Azevedo C, Ramalho NM & Knoppers B, (1994) A biomassa de *Chara hornemannii* em relação ao comportamento físico-químico da lagoa de Piratininga (RJ). Anais da Academia Brasileira de Ciências 66(2): 213-222.

[9] Chambers PA & Kalff J (1985) Depth distribution and biomass of submersed aquatic macrophyte communities in relation to Secchi depth. Canadian Journal of Fisheries and Aquatic Sciences 42: 701-709.

[10] Esteves FA (Coord.) (2011) Fundamentos de Limnologia. Editora Interciência, Rio de Janeiro. (3rd edition).

[11] Esteves FA & Camargo AFM (1986) Caracterização de sedimentos de 17 reservatórios do Estado de São Paulo com base no teor de feopigmentos, carbono orgânico e nitrogênio orgânico. Ciência e Cultura 34(5): 669-674.

[12] Fernández-Aláez M, Fernández-Aláez C & Bécares E (1999) Nutrient contents in macrophytes in Spanish shallow lakes. Hydrobiologia 408-409: 317-326.

[13] Fernández-Aláez M, Fernández-Aláez C & Rodríguez S (2002) Seasonal changes in biomass of charophytes in shallow lakes northwest of Spain. Aquatic Botany 72(3-4): 335-348.

[14] Howard-Williams C, Schwarz AM & Vincent WF (1995) Deep-water aquatic plant communities in an oligotrophic lake: physiological responses do variable light. Freshwater Biology 33: 91-102.

[15] Hunter RD (1976) Changes in carbon and nitrogen content during decomposition of three macrophytes in freshwater and marine environments. Hydrobiologia 51(2): 119-128.

[16] Kosten S, Jeppesen L, Huszar VLM, Mazzeo N, van Nee EH, Peeters ETHM & Lürling M (2011) Ambiguous climate impacts on competition between submerged macrophytes and phytoplankton in shallow lakes. Freshwater Biology 56: 1540-1553.

[17] Krause W (1985) Über die Standortsansprüche und das Ausbreitungs-verhalten der Stern-Armleucheralge Nitellopsis obtusa (Desvaux) J. Groves. Carolinea 42(4): 31-42.

[18] Krebs CJ (1989) Ecological methodology. Harper & Row Publishers, New York.

[19] Kufel I & Kufel L (1997) Eutrophication processes in a shallow macrophyte-dominated lake: nutrient loading to and flow through lake Luknajno (Poland). Hydrobiologia 342-343: 387-394.

[20] Kufel L & Kufel I (2002) Chara beds acting as nutrient sinks in shallow lakes: a review. Aquatic Botany 72: 249-260.

[21] Murphy J & Riley J (1962) A modified single solution method for determination of phosphate in natural waters. Analytica Chemica Acta 27: 31-36.

[22] Nõges P, Tuvikene L, Feldmann T, Tõnno I, Künnap H, Luup H, Salujõe J & Nõges T (2003) The role of charophytes in increasing water transparency: a case study of two shallow lakes in Estonia. Hydrobiologia 506 509: 567-573.

[23] Palma-Silva C, Albertoni EF & Esteves FA (2002) Clear water associated with biomass and nutrient variation during the growth of a charophyte stand after a drawdown, in a tropical coastal lagoon. Hydrobiologia 482: 79-87.

[24] Palmer M & Reid BR (2010) Biomass production and processing and methods of use thereof. Patentdocs: patent application number 20100240114. Available: http://www.faqs.org/patents/app/20100240114.

[25] Pereyra-Ramos E (1981) The ecological role of Characeae in the lake littoral. Ekologia Polska 29(2): 167-209.

[26] Rich G & Cherry K (1987) Hazardous waste treatment technologies. Pudvan Publishers, New York.

[27] Rip WJ, Ouboter MRL & Los HJ (2007) Impact of climatic fluctuations on Characeae biomass in a shallow, restored lake in The Netherlands. Hydrobiologia 584: 415-424.

[28] Rodusky AJ, Sharfstein B, East TL & Maki RP (2005) A comparison of three methods to collect submerged aquatic vegetation in a shallow lake. Environmental Monitoring and Assessment 110: 87-97.

[29] Salmaso, N, Buzzi F, Garibaldi L, Morabito G & Simona M (2012) Effects of nutrient availability and temperature on phytoplankton development: a case study of large lakes south of the Alps. Aquatic Science (on line first).

[30] Scheffer M & van Nes EH (2007) Shallow lakes theory revisited: various alternative regimes driven by climate, nutrients and lake size. Hydrobiologia 584: 455-466.

[31] Schwarz AM & Hawes I (1997) Effects of changing water clarity on characean biomass and species composition in a large oligotrophic lake. Aquatic Botany 56: 69-181.

[32] Sender J (2008) Long term changes of macrophytes structure in the lake Moszne (Poleski National Park). Teka Komisji Ochromy i Ksztaltowania Srodowiska Pryzyrodniczego 5: 154-163.

[33] Shaikh Parveen R & Bhosle Arjun B (2011) Bioaccumulation of chromium by aquatic macrophytes Hydrilla sp. & Chara sp. Advances in Applied Sciences Research 21(1): 214-220.

[34] Siong K & Asaeda T (2009) Effect of magnesium on charophytes calcification: implication for phosphorus speciation stored in biomass and sediment in Myall Lake (Australia). Hydrobiologia 632: 247-259.

[35] Siong K, Asaeda T, Fujino T & Redden A (2006) Difference characteristics of phosphorus in Chara and two submerged angiosperm species: implications for phosphorus nutrient cycling in an aquatic ecosystem. Wetlands Ecology and Management 14: 505-510.

[36] Stephen D, Moss B & Phillips G (1998) The relative importance of top-down and bottom-up control of phytoplankton in a shallow macrophyte-dominated lake. Freshwater Biology 39: 699-713.

[37] Steinman AD, Meeker RH, Rodusky AJ, Davis WP & Hwang S-J (1997) Ecological properties of charophytes in a large subtropical lake. Journal of the North American Benthological Society 16(4): 781-793.

[38] Strickland JDH & Parsons TR (1965) A manual of sea water analysis. Bulletin of the Fisheries Research Board of Canada 125: 1-185.

[39] Talling JF & Parker JE (2002) Seasonal dynamics of phytoplankton and phytobenthos, and associated chemical interactions, in a shallow upland lake (Malham Tarn, northern England). Hydrobiologia 487: 167-181.

[40] Torn K, Martin G & Paalme T (2006) Seasonal changes in biomass, elongation growth and primary production rate of Chara tomentosa in the NE Baltic Sea. Annales Botanici Fennici 43: 276-283.

[41] van de Bund WJ & van Donk E (2004) Effects of fish and nutrient additions on food-web stability of a charophyte-dominated lake. Freshwater Biology 49(12): 1565-1573.

[42] Westlake DF (1965) Some basic data for investigations of the productivity of aquatic macrophytes. Memorie dell'Istituto Italiano di Idrobiologia Dottore Marco de Marchi 18: 229-248.

[43] Westlake DF (1971) Macrophytes, In: Vollenweider RA. A manual of methods for measuring primary production in aquatic environments. Blackwell Scientific Publications, Oxford (IBP Handbook 12). p. 25-32 (9th edition).

[44] Wetzel RG (1964) A comparative study of the primary productivity of higher aquatic plants, periphyton and phytoplankton in a large, shallow lake. Internationale Revue der Gesamten Hydrobiologie 49(1): 1-61.

[45] Wetzel RG (2001) Limnology: lake and river ecosystems. Academic Press, Orlando, Florida (3rd edition).

Permissions

The contributors of this book come from diverse backgrounds, making this book a truly international effort. This book will bring forth new frontiers with its revolutionizing research information and detailed analysis of the nascent developments around the world.

We would like to thank Miodrag Darko Matovic, for lending his expertise to make the book truly unique. He has played a crucial role in the development of this book. Without his invaluable contribution this book wouldn't have been possible. He has made vital efforts to compile up to date information on the varied aspects of this subject to make this book a valuable addition to the collection of many professionals and students.

This book was conceptualized with the vision of imparting up-to-date information and advanced data in this field. To ensure the same, a matchless editorial board was set up. Every individual on the board went through rigorous rounds of assessment to prove their worth. After which they invested a large part of their time researching and compiling the most relevant data for our readers. Conferences and sessions were held from time to time between the editorial board and the contributing authors to present the data in the most comprehensible form. The editorial team has worked tirelessly to provide valuable and valid information to help people across the globe.

Every chapter published in this book has been scrutinized by our experts. Their significance has been extensively debated. The topics covered herein carry significant findings which will fuel the growth of the discipline. They may even be implemented as practical applications or may be referred to as a beginning point for another development. Chapters in this book were first published by InTech; hereby published with permission under the Creative Commons Attribution License or equivalent.

The editorial board has been involved in producing this book since its inception. They have spent rigorous hours researching and exploring the diverse topics which have resulted in the successful publishing of this book. They have passed on their knowledge of decades through this book. To expedite this challenging task, the publisher supported the team at every step. A small team of assistant editors was also appointed to further simplify the editing procedure and attain best results for the readers.

Our editorial team has been hand-picked from every corner of the world. Their multi-ethnicity adds dynamic inputs to the discussions which result in innovative outcomes. These outcomes are then further discussed with the researchers and contributors who give their valuable feedback and opinion regarding the same. The feedback is then collaborated with the researches and they are edited in a comprehensive manner to aid the understanding of the subject.

Apart from the editorial board, the designing team has also invested a significant amount of their time in understanding the subject and creating the most relevant covers. They scrutinized every image to scout for the most suitable representation of the subject and create an appropriate cover for the book.

The publishing team has been involved in this book since its early stages. They were actively engaged in every process, be it collecting the data, connecting with the contributors or procuring relevant information. The team has been an ardent support to the editorial, designing and production team. Their endless efforts to recruit the best for this project, has resulted in the accomplishment of this book. They are a veteran in the field of academics and their pool of knowledge is as vast as their experience in printing. Their expertise and guidance has proved useful at every step. Their uncompromising quality standards have made this book an exceptional effort. Their encouragement from time to time has been an inspiration for everyone.

The publisher and the editorial board hope that this book will prove to be a valuable piece of knowledge for researchers, students, practitioners and scholars across the globe.

List of Contributors

Adina-Elena Segneanu, Paulina Vlazan, Paula Sfirloaga and Ioan Grozescu
National Institute of Research and Development for Electrochemistry and Condensed Matter – INCEMC Timisoara, Romania

Florentina Cziple
Eftimie Murgu University, Resita, Romania

Vasile Daniel Gherman
Politehnica University of Timisoara, Romania

Hongbin Cheng
Department of Process Engineering, Stellenbosch University, South Africa
New China Times Technology Ltd, China

Lei Wang
Department of Life Science, Imperial College London, UK
New China Times Technology Ltd, China

Shurong Wang
Zhejiang University, China

K.L. Chin and P.S. H'ng
Faculty of Forestry, Universiti Putra Malaysia, UPM Serdang, Selangor, Malaysia

Ernesto A. Chávez and Alejandra Chávez-Hidalgo
Centro Interdisciplinario de Ciencias Marinas, Instituto Politécnico Nacional, La Paz, México

Onofre Monge Amaya, María Teresa Certucha Barragán and Francisco Javier Almendariz Tapia
University of Sonora, Department of Chemistry and Metallurgy, Hermosillo, Sonora, México

Werther Guidi, Frédéric E. Pitre and Michel Labrecque
Institut de Recherche en Biologie Végétale (IRBV – Plant Biology Research Institute) – Université de Montréal – The Montreal Botanical Garden, Montréal, Canada

Theo Verwijst, Anneli Lundkvist and Stina Edelfeldt
Department of Crop Production Ecology, Swedish University of Agricultural Sciences, Uppsala, Sweden

Johannes Albertsson
Department of Plant Breeding and Biotechnology, Swedish University of Agricultural Sciences, Alnarp, Sweden

Carlos E. de M. Bicudo
Instituto de Botânica, São Paulo, SP, Brasil

Norma C. Bueno
Universidade Estadual do Oeste do Paraná, Cascavel, PR, Brasil